CROM
KE

Australian Reptiles & Frogs

This edition published in 2014

First published in 2009

Jacana Books, an imprint of
Allen & Unwin
83 Alexander Street
Crows Nest NSW 2065
Australia
Phone: (61 2) 8425 0100
Email: info@allenandunwin.com
Web: www.allenandunwin.com

Cataloguing-in-Publication details are available
from the National Library of Australia
www.trove.nla.gov.au

ISBN 978 1 76011 105 2

A LEONARD CRONIN PRODUCTION
Text: Leonard Cronin
Illustrations: Nicola Oram, Barbara Duckworth, Mike Gorman, Marion Westmacott
Distribution Maps: Laurel Cohn
Design: Robert Taylor
Editor: Laurel Cohn

Printed by HangTai Printing Company Limited, China

10 9 8 7 6 5 4 3 2 1

CRONIN'S
KEY GUIDE

Australian Reptiles & Frogs

Leonard Cronin

JACANA BOOKS

ALLEN&UNWIN

Strophurus taenicauda
Golden-tailed Gecko

Contents

Introduction

Australia is home to some 917 species of reptiles and 227 frogs. Among them are the largest and most fearsome predator on earth — the estuarine crocodile; the world's deadliest snake — the mainland taipan; a lizard that walks on water — the beaded gecko; and an amphibian from the harshest deserts — the water-holding frog. Newts, salamanders and caecilians (worm-like, burrowing amphibians) are absent from our amphibious fauna, and Australia's only true toad is the introduced cane toad.

Among the reptiles are snakes, lizards, crocodiles and turtles. They differ widely in body form, but all have scaly skin that minimises water loss, and many are able to produce near-solid urine, enabling them to populate the arid, inland areas. Australia's reptiles have developed a range of adaptations to cope with drought and heat. The scales of the thorny devil, for example, soak up water from night-time dew and moist sand, and channel it to the corner of the lizard's mouth by capillary action. Most desert-dwelling reptiles hide from the heat in a burrow, and only emerge at night to feed. Burrows, however, indicate to a predator the whereabouts of a potential meal. To avoid being trapped some reptiles dig complex burrow systems with ventilation shafts and escape routes. Others plug the entrances or conceal them beneath vegetation, and some skinks inflate their body with air and jam themselves in the burrow. Their spiny scales grip the sides of the tunnel, making them almost impossible to pull out.

Many reptiles spend part or all of their lives in trees, and a number of geckos have enlarged toes with hundreds of tiny overlapping plates on the soles. These act like suction pads allowing the gecko to grip smooth surfaces, climb vertically and run upside down.

Reptiles and amphibians do not produce enough heat to maintain a constant body temperature, and are referred to as ectothermic or cold-blooded. Being ectothermic confines most of the reptiles and amphibians to warmer climates, and the number of species declines as the altitude and latitude increase. Ectothermic animals have the advantage, however, of needing much less food than the endothermic or warm-blooded birds and mammals. When it is too cold for them to function, reptiles and amphibians slow down their metabolism and enter a state of torpor, enabling them to survive for long periods without food or water.

In cool conditions reptiles are sluggish and often warm themselves in the morning sun so that they can function at full speed. They are particularly vulnerable to predators at this time, and most lizards are able to shed all or part of their tail as a sacrifice, and regenerate a new one. Some, like the frilled lizard, rely on aggressive displays to deter predators, backed up by biting or tail swiping, and many snakes can inject toxic venom. A number of geckos smear their attacker with a thick, sticky and irritating fluid secreted from glands in the tail.

Reptiles (except the Tuatara of New Zealand) fertilise their eggs internally, and males are equipped with a penis, or two in the case of snakes. Females typically lay leathery shelled eggs in a burrow or crevice where they are left to hatch on their own, although some species, generally in cold climates, give birth to live young. Most offer no parental care, and few offspring survive to adulthood. Some female pythons, however, coil around their eggs to incubate and protect them, shivering to increase their body temperature on cold days. Female crocodiles also guard their eggs and often carry hatchlings to the water in their mouths.

Limbless reptiles

Snakes and legless lizards grip the substrate with their overlapping belly scales and move forward by waves of muscular contraction. This method of locomotion is surprisingly efficient, and whip snakes can move at 10 kph.

Snakes are, in effect, long tubes, and this body shape limits the size of the mouth in relation to the rest of the body, and hence the size of food items that can be eaten. To overcome this limitation the two halves of the snake's jaw are joined by an elastic ligament and are able to separate widely, allowing them to swallow prey much larger than their heads. Snakes can consume prodigious meals and survive for weeks or months without feeding.

Snakes rely mainly on scent to detect prey, and like the monitors have a long, deeply forked tongue that flicks in and out, gathering chemicals from the environment and transferring them to the Jacobson's organ in the roof of the mouth. Like our organs of taste and smell, Jacobson's organ gives the snake information about the presence of other animals in its immediate surroundings. Although snakes have no external ear openings or ear drums, they are not deaf, and are able to detect tiny vibrations through the ground or water. Most Australian pythons detect warm-blooded prey using heat-sensitive organs located in pits in the scales bordering the lips.

Amphibian adaptations

Australia is the driest habitable continent, and many of our frogs have adapted their life cycles, behaviour and physiology to cope with extended droughts and short wet seasons. An occasional inflow of water, either from the sky or from temporary waterways, is usually sufficient to support a frog population.

Desert-dwelling frogs cope with droughts by hiding in soil cracks or burying themselves beneath drying ponds. There, in a small chamber up to 50 cm underground, they enclose themselves in a cocoon made by shedding several layers of skin to form a waterproof sack, leaving only their nostrils exposed. With a bladder full of water they can survive for many months underground, slowing down their metabolism and digging themselves out when surface water penetrates their burial chamber.

Most frogs prefer warm, moist habitats, and in the coastal forests many species have taken to the trees. Tree frogs have grooved discs or pads on the tips of their fingers and toes that allow them to 'stick' onto leaves and other smooth, vertical surfaces. In the dry season they hide in moist leaf bases to prevent dehydration.

The life cycle of frogs is unusual among the land vertebrates. The eggs are fertilised outside the body by a cloud of sperm deposited by the male over the eggs. The eggs are generally laid in water where they hatch into completely aquatic, free-swimming, fish-like tadpoles that breathe through gills. After a few weeks feeding on algae and other aquatic vegetation, the tadpoles begin to grow legs and gradually transform into small frogs. These tiny froglets emerge from the water, breathe air through their lungs and are generally land-dwellers, although many spend a great deal of time in the water.

All frogs are carnivores, eating virtually any small animal that will fit into their mouth. Most are, however, incapable of swallowing anything larger than insects, other arthropods and earthworms. Larger frogs, like the 10 cm long north-eastern water-holding frog, occasionally eat small reptiles and other frogs. Prey is detected by sight and is typically caught with the frog's long, muscular, sticky tongue.

Frogs absorb water through the skin on their undersides and never need to drink. Most can change colour within a few hours or days to absorb or reflect heat or to match their surroundings and avoid being seen by predators. Some, like the crucifix toad, can secrete poisonous fluids from glands in their skin to deter predators. Frogs are more likely to be heard than seen, and their distinctive calls are a useful aid to identification. Most are nocturnal, and the repertoire of thousands of calling males can create a deafening chorus on moist summer evenings.

IDENTIFYING SNAKES

Many snake species exhibit a wide range of colours, making them very difficult to identify. Each species does, however, have a similar number of scales around the middle of their body, providing an invaluable aid to snake identification. Scales are counted in a row, and are easily seen on a shed skin. If you are able to observe the scales on a living snake, count the rows on one side, from the belly to the mid-back, excluding the large belly scales, and double the number. Do not attempt to pick up a snake as this could prove deadly. The diagram shows the scales on a flattened section of snake skin.

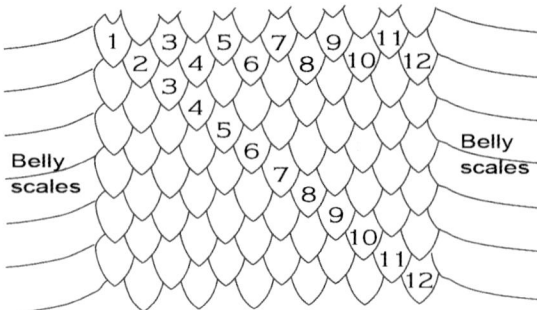

SNAKE BITE

Snakes generally try to avoid humans and most snake bites are the result of someone attempting to kill or handle a snake. If you come across a snake, the best course of action is to freeze and then slowly back away. Snakes will usually tolerate being momentarily stepped on, but to minimise the risk of being accidentally bitten, you should wear long trousers and boots when walking through long grass. Carry a torch at night and keep areas around houses clean, remove rubbish and cut the grass. Bushwalkers should take a long, broad crepe bandage with them to use in an emergency.

There are around 3000 snake bites per year in Australia. Fewer than 500 are treated with antivenene, and on average only one or two bites will prove fatal. Brown snakes are responsible for about half the deaths, followed by tiger snakes, taipans and death adders. It is, however, uncommon to die within 4 hours of being bitten, and although death from snake bite is very rare, all bites should be treated immediately and the victim taken to hospital, even if you think the snake was not poisonous or if you cannot see 2 puncture wounds.

First Aid

• Do not cut the wound. Do not apply an arterial torniquet. Do not attempt to remove the venom by washing or sucking. Traces of venom are used by the hospital to identify the snake.

• Apply a broad constrictive bandage as soon as possible, working from the bite towards the heart. Wind the bandage as firmly as you would for a sprained ankle but not so tight that blood circulation is prevented. Extend the bandage over the whole limb if possible. If you do not have a bandage use a piece of clothing or other material.

• If the bite is not on a limb, apply and maintain firm pressure to the site of the bite to prevent distribution of the venom.

• Immobilise the limb using a stick, rolled newspaper or a sling.

• Reassure the patient and do not give alcohol. Keep the patient still and telephone for help if you can. Take the patient to the nearest hospital as soon as possible. Transport should be taken to the patient, but if that is impractical, carry the patient, or if there is no other alternative the patient should walk but not run.

• Do not remove the bandage until the patient is in a medical treatment area.

• Do not attempt to kill the snake for identification as hospitals have kits to determine the type of venom.

HOW TO USE THIS GUIDE

The reptiles and amphibians in this book have been carefully selected to represent the different families, and include the most commonly encountered species and those which are likely to be noticed in their habitats.

The visual key on the following pages includes miniatures of all the species in the book with their measurements from snout to tail tip, the page number and a distribution map. Bear in mind that different populations of the same species may show differences in colour and pattern. There may also be colour variations in juveniles and different sexes, and many reptiles and frogs have the ability to change colour depending on their surroundings and the temperature.

The distribution maps indicate areas where an animal may be found, although this depends on the presence of suitable habitat with sufficient food, water and shelter. The distribution of a species is usually quite discontinuous within its range, and the blank areas on the map showing where a species is not found are often more informative than the coloured areas.

Status in the text shows the threat status of a species according to the Red List of Threatened Species compiled by the International Union for Conservation of Nature. In some cases, however, a subspecies may be at risk, although the species as a whole is not. If a species is not on the Red List I have used the term low risk, or included any State classifications.

According to the Commonwealth (2009) there are 7 critically endangered, 17 endangered and 34 vulnerable reptiles in Australia, and 5 critically endangered, 14 endangered and 10 vulnerable frogs. These shocking statistics make Australia home to more threatened reptiles and amphibians than any other country on Earth. Land clearing and its consequences, such as salinisation of rivers and landscapes, are the foremost threat to the majority of species on this list. Land clearing throughout Australia must be stopped and a massive revegetation program established to compensate for the damage already done.

FROGS

Length
to **45 mm**

Tusked Frog **26**

Length
to **70 mm**

Eastern Banjo Frog **31**

Length
to **80 mm**

Stuttering Frog **36**

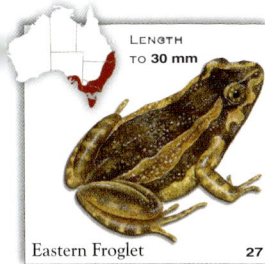

Length
to **30 mm**

Eastern Froglet **27**

Length
to **42 mm**

Ornate Burrowing Frog **32**

Length
to **80 mm**

Great Barred Frog **37**

Length
to **90 mm**

Giant Burrowing Frog **28**

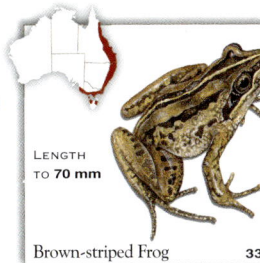

Length
to **70 mm**

Brown-striped Frog **33**

Length
to **60 mm**

Turtle Frog **38**

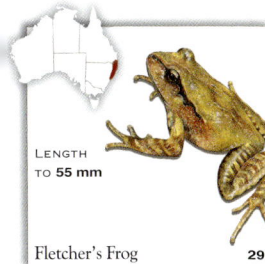

Length
to **55 mm**

Fletcher's Frog **29**

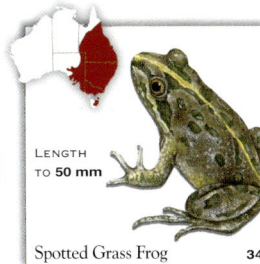

Length
to **50 mm**

Spotted Grass Frog **34**

Length
to **40 mm**

Painted Burrowing Frog **39**

Length
to **55 mm**

Marbled Frog **30**

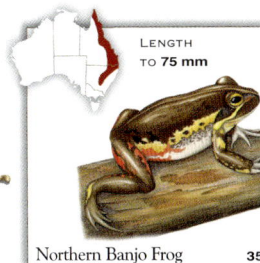

Length
to **75 mm**

Northern Banjo Frog **35**

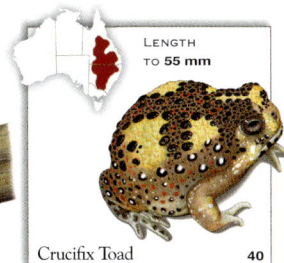

Length
to **55 mm**

Crucifix Toad **40**

FROGS

LENGTH
TO **35 mm**

Haswell's Froglet **41**

LENGTH
TO **70 mm**

Striped Burrowing Frog **46**

LENGTH
TO **85 mm**

Green and Golden Bell Frog **51**

LENGTH
TO **30 mm**

Red-crowned Toadlet **42**

LENGTH
TO **100 mm**

Giant Frog **47**

LENGTH
TO **30 mm**

Northern Dwarf Tree Frog **52**

LENGTH
TO **35 mm**

Red-backed Toadlet **43**

LENGTH
TO **45 mm**

Short-footed Frog **48**

LENGTH
TO **100 mm**

Green Tree Frog **53**

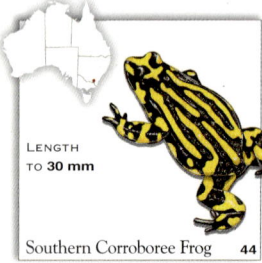

LENGTH
TO **30 mm**

Southern Corroboree Frog **44**

LENGTH
TO **100 mm**

New Holland Frog **49**

LENGTH
TO **65 mm**

Red-eyed Green Tree Frog **54**

LENGTH
TO **32 mm**

Smooth Toadlet **45**

LENGTH
TO **72 mm**

Water-holding Frog **50**

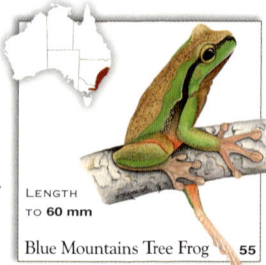

LENGTH
TO **60 mm**

Blue Mountains Tree Frog **55**

FROGS

LENGTH
TO **70 mm**

Dahl's Aquatic Frog   **56**

LENGTH
TO **140 mm**

Giant Tree Frog   **61**

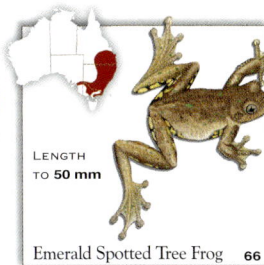

LENGTH
TO **50 mm**

Emerald Spotted Tree Frog   **66**

LENGTH
TO **45 mm**

Bleating Tree Frog   **57**

LENGTH
TO **40 mm**

Broad-palmed Rocket Frog   **62**

LENGTH
TO **40 mm**

Leaf Green Tree Frog   **67**

LENGTH
TO **25 mm**

Eastern Dwarf Tree Frog   **58**

LENGTH
TO **70 mm**

Stony Creek Frog   **63**

LENGTH
TO **105 mm**

Growling Grass Frog   **68**

LENGTH
TO **42 mm**

Freycinet's Rocket Frog   **59**

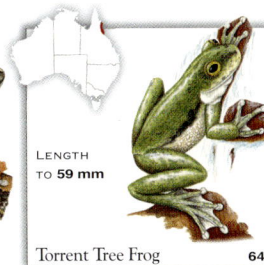

LENGTH
TO **59 mm**

Torrent Tree Frog   **64**

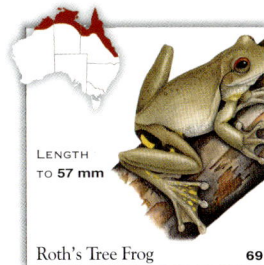

LENGTH
TO **57 mm**

Roth's Tree Frog   **69**

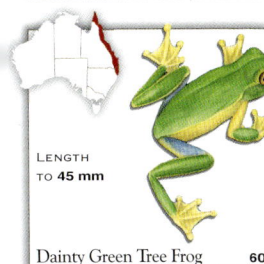

LENGTH
TO **45 mm**

Dainty Green Tree Frog   **60**

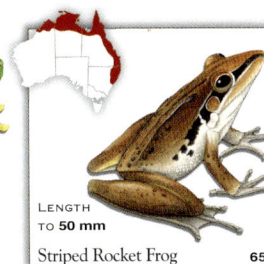

LENGTH
TO **50 mm**

Striped Rocket Frog   **65**

LENGTH
TO **35 mm**

Desert Tree Frog   **70**

FROGS

LENGTH
TO 120 mm

Magnificent Tree Frog 71

LENGTH
TO 35 mm

Whistling Tree Frog 72

LENGTH
TO 81 mm

Wood Frog 73

GECKOS

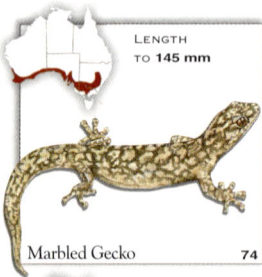

LENGTH
TO 145 mm

Marbled Gecko 74

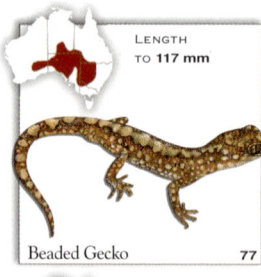

LENGTH
TO 117 mm

Beaded Gecko 77

LENGTH
TO 85 mm

Wood Gecko 80

LENGTH
TO 370 mm

Ring-tailed Gecko 75

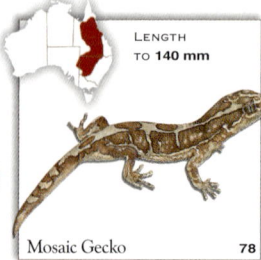

LENGTH
TO 140 mm

Mosaic Gecko 78

LENGTH
TO 145 mm

House Gecko 81

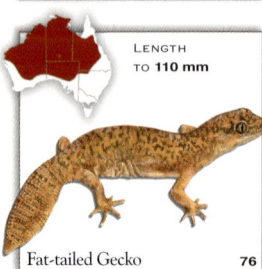

LENGTH
TO 110 mm

Fat-tailed Gecko 76

LENGTH
TO 90 mm

Crowned Gecko 79

LENGTH
TO 133 mm

Tree Dtella 82

GECKOS

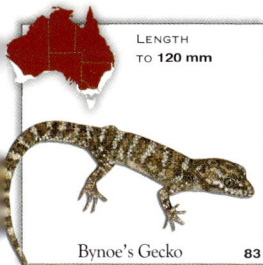

LENGTH
TO **120 mm**

Bynoe's Gecko   **83**

LENGTH
TO **164 mm**

Robust Velvet Gecko   **88**

LENGTH
TO **170 mm**

Spiny-tailed Gecko   **93**

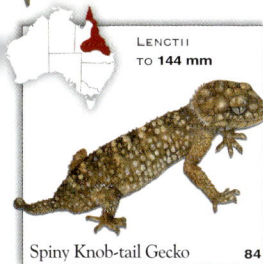

LENGTH
TO **144 mm**

Spiny Knob-tail Gecko   **84**

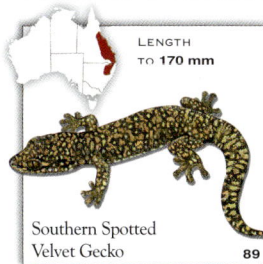

LENGTH
TO **170 mm**

Southern Spotted
Velvet Gecko   **89**

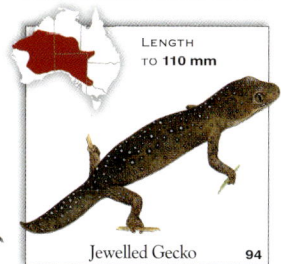

LENGTH
TO **110 mm**

Jewelled Gecko   **94**

LENGTH
TO **115 mm**

Smooth Knob-tail Gecko   **85**

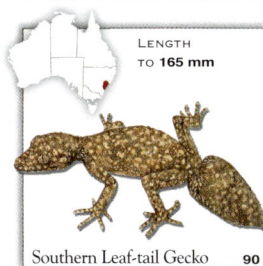

LENGTH
TO **165 mm**

Southern Leaf-tail Gecko   **90**

LENGTH
TO **120 mm**

Golden-tailed Gecko   **95**

LENGTH
TO **170 mm**

Northern Velvet Gecko   **86**

LENGTH
TO **103 mm**

Beaked Gecko   **91**

LENGTH
TO **142 mm**

Eastern Spiny-tailed Gecko   **96**

LENGTH
TO **130 mm**

Lesueur's Velvet Gecko   **87**

LENGTH
TO **224 mm**

Northern Leaf-tail Gecko   **92**

LENGTH
TO **156 mm**

Barking Gecko   **97**

DRAGONS

LENGTH
TO **350 mm**

Jacky Dragon 98

LENGTH
TO **190 mm**

Painted Dragon 103

LENGTH
TO **425 mm**

Gilbert's Dragon 107

LENGTH
TO **320 mm**

Chameleon Dragon 99

LENGTH
TO **300 mm**

Two-lined Dragon 104

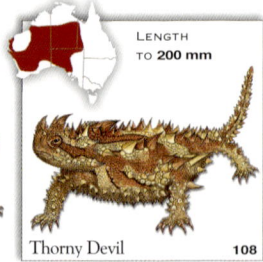

LENGTH
TO **200 mm**

Thorny Devil 108

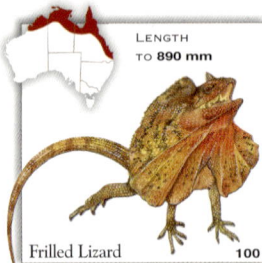

LENGTH
TO **890 mm**

Frilled Lizard 100

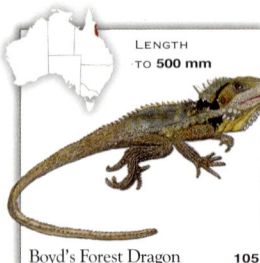

LENGTH
TO **500 mm**

Boyd's Forest Dragon 105

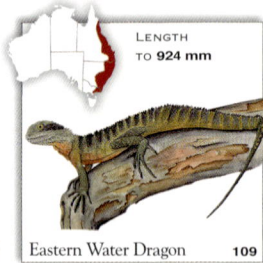

LENGTH
TO **924 mm**

Eastern Water Dragon 109

LENGTH
TO **350 mm**

Ring-tailed
Dragon 101

LENGTH
TO **370 mm**

Southern
Angle-headed Dragon 106

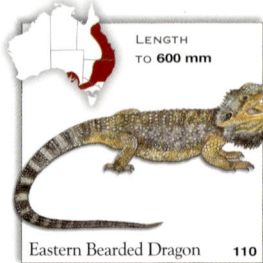

LENGTH
TO **600 mm**

Eastern Bearded Dragon 110

LENGTH
TO **265 mm**

Central Netted Dragon 102

LENGTH
TO **990 mm**

Mountain Dragon 111

MONITORS

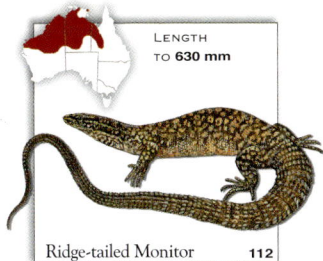

LENGTH
TO **630 mm**

Ridge-tailed Monitor 112

LENGTH
TO **1.3 m**

Merten's Water Monitor 117

LENGTH
TO **2.5 m**

Perentie 113

LENGTH
TO **680 mm**

Mitchell's Water Monitor 118

LENGTH
TO **220 mm**

Pygmy Mulga Monitor 114

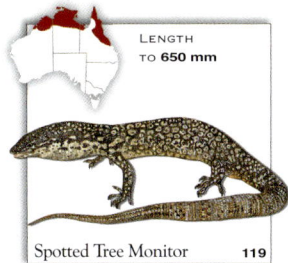

LENGTH
TO **650 mm**

Spotted Tree Monitor 119

LENGTH
TO **1.6 m**

Gould's Goanna 115

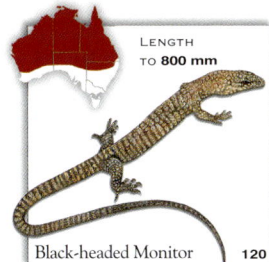

LENGTH
TO **800 mm**

Black-headed Monitor 120

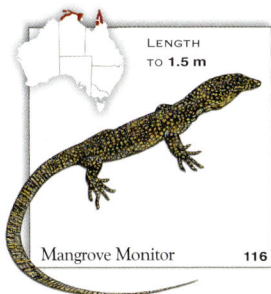

LENGTH
TO **1.5 m**

Mangrove Monitor 116

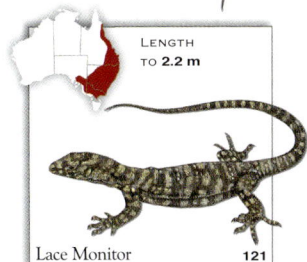

LENGTH
TO **2.2 m**

Lace Monitor 121

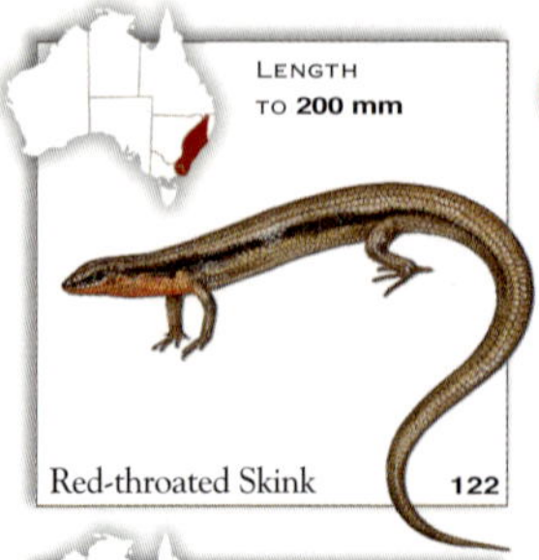

SKINKS

LENGTH
TO **200 mm**

Red-throated Skink   **122**

LENGTH
TO **326 mm**

Leopard Skink   **127**

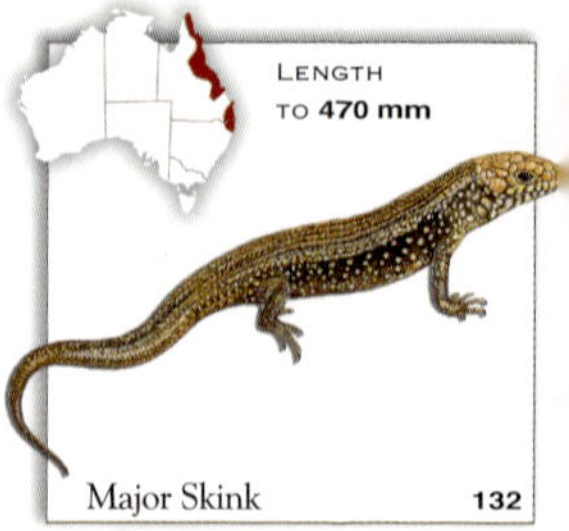

LENGTH
TO **470 mm**

Major Skink   **132**

LENGTH
TO **150 mm**

Blue-throated Rainbow Skink **123**

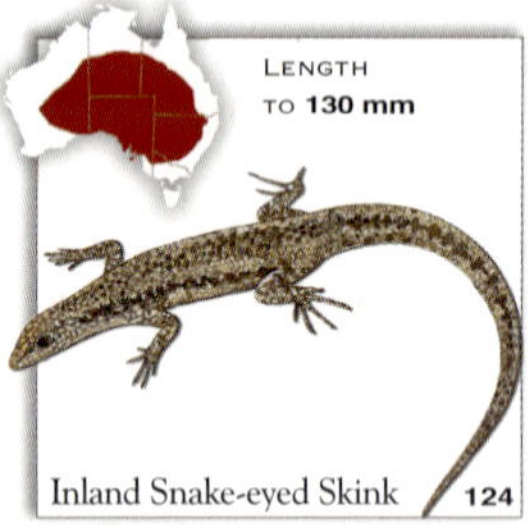

LENGTH
TO **250 mm**

Eastern Striped Skink   **128**

LENGTH
TO **200 mm**

Desert Skink   **133**

LENGTH
TO **130 mm**

Inland Snake-eyed Skink   **124**

LENGTH
TO **200 mm**

Copper-tailed Skink   **129**

LENGTH
TO **750 mm**

Land Mullet   **134**

LENGTH
TO **100 mm**

Wall Skink   **125**

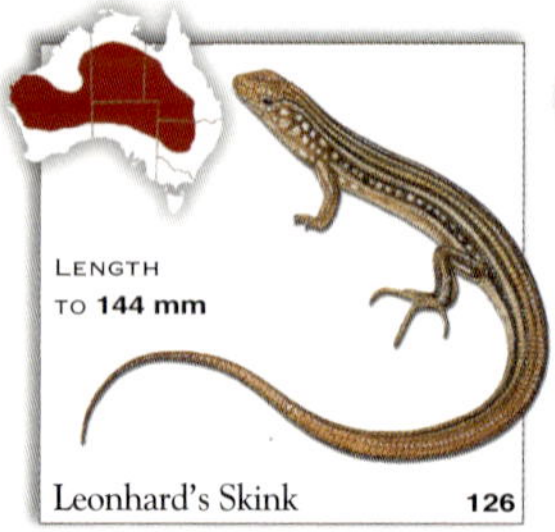

LENGTH
TO **450 mm**

Cunningham's Skink   **130**

LENGTH
TO **285 mm**

Gidgee Skink   **135**

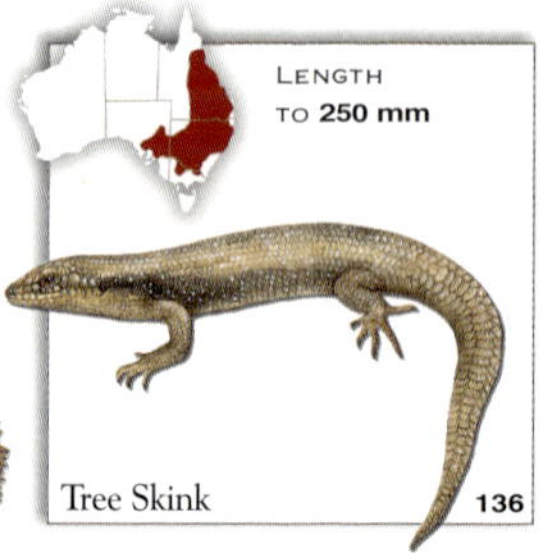

LENGTH
TO **144 mm**

Leonhard's Skink   **126**

LENGTH
TO **161 mm**

Pygmy Spiny-tailed Skink   **131**

LENGTH
TO **250 mm**

Tree Skink   **136**

SKINKS

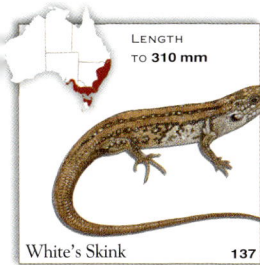

LENGTH
TO **310 mm**

White's Skink   **137**

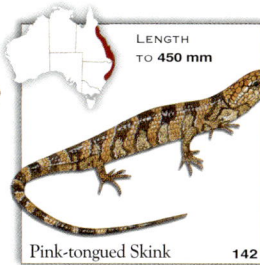

LENGTH
TO **450 mm**

Pink-tongued Skink   **142**

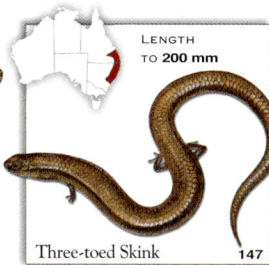

LENGTH
TO **200 mm**

Three-toed Skink   **147**

LENGTH
TO **300 mm**

Broad-banded Sand Swimmer **138**

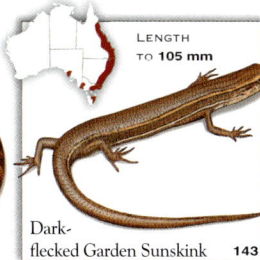

LENGTH
TO **105 mm**

Dark-
flecked Garden Sunskink   **143**

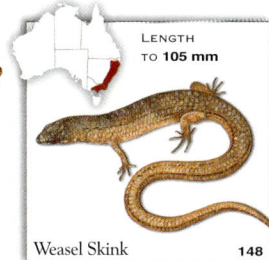

LENGTH
TO **105 mm**

Weasel Skink   **148**

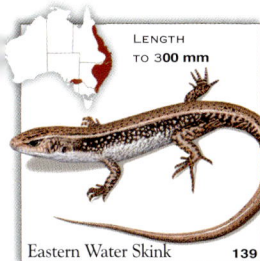

LENGTH
TO **300 mm**

Eastern Water Skink   **139**

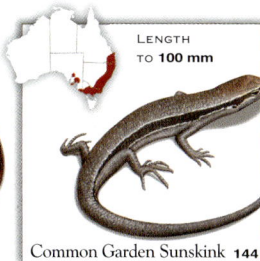

LENGTH
TO **100 mm**

Common Garden Sunskink   **144**

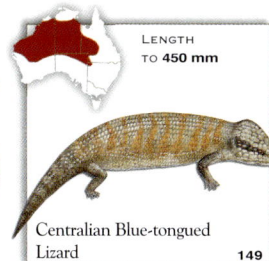

LENGTH
TO **450 mm**

Centralian Blue-tongued
Lizard   **149**

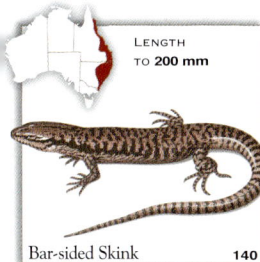

LENGTH
TO **200 mm**

Bar-sided Skink   **140**

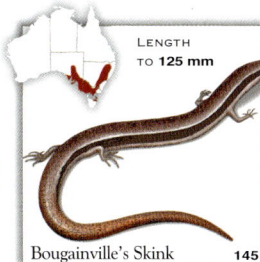

LENGTH
TO **125 mm**

Bougainville's Skink   **145**

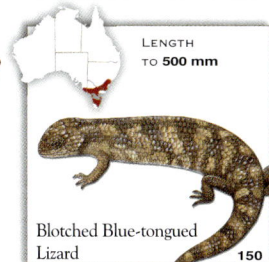

LENGTH
TO **500 mm**

Blotched Blue-tongued
Lizard   **150**

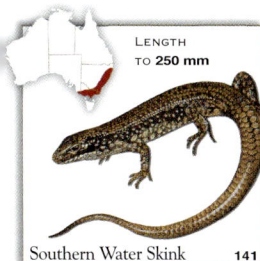

LENGTH
TO **250 mm**

Southern Water Skink   **141**

LENGTH
TO **90 mm**

Lined Fire-tailed Skink   **146**

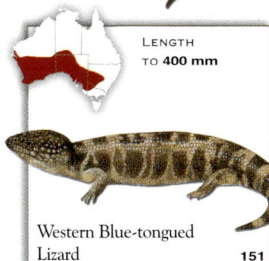

LENGTH
TO **400 mm**

Western Blue-tongued
Lizard   **151**

SKINKS

LENGTH
TO **420 mm**

Shingleback Lizard **152**

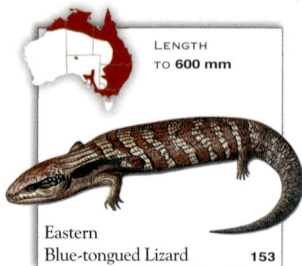

LENGTH
TO **600 mm**

Eastern
Blue-tongued Lizard **153**

LEGLESS LIZARDS

LENGTH
TO **267 mm**

Excitable Delma **154**

LENGTH
TO **840 mm**

Common Scaly Foot **156**

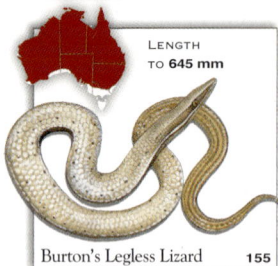

LENGTH
TO **645 mm**

Burton's Legless Lizard **155**

LENGTH
TO **575 mm**

Western Hooded Scaly Foot **157**

SNAKES

LENGTH
TO **750 mm**

Common
Eastern Blind Snake   **158**

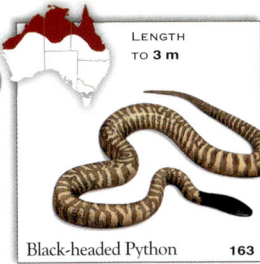

LENGTH
TO **3 m**

Black-headed Python   **163**

LENGTH
TO **4 m**

Carpet Python   **168**

LENGTH
TO **320 mm**

Brown-snouted Blind Snake   **159**

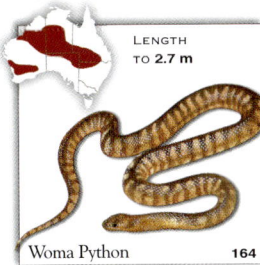

LENGTH
TO **2.7 m**

Woma Python   **164**

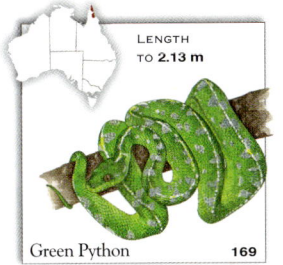

LENGTH
TO **2.13 m**

Green Python   **169**

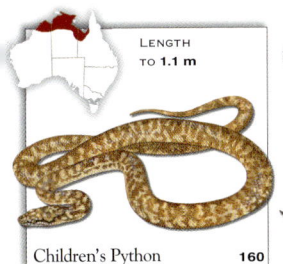

LENGTH
TO **1.1 m**

Children's Python   **160**

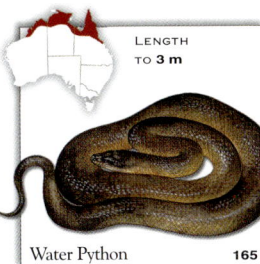

LENGTH
TO **3 m**

Water Python   **165**

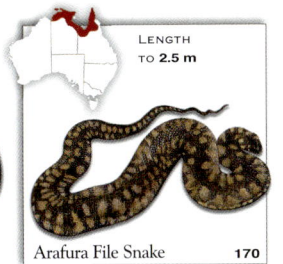

LENGTH
TO **2.5 m**

Arafura File Snake   **170**

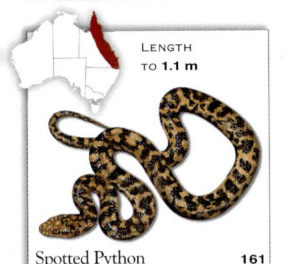

LENGTH
TO **1.1 m**

Spotted Python   **161**

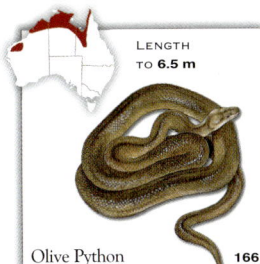

LENGTH
TO **6.5 m**

Olive Python   **166**

LENGTH
TO **2 m**

Brown Tree Snake   **171**

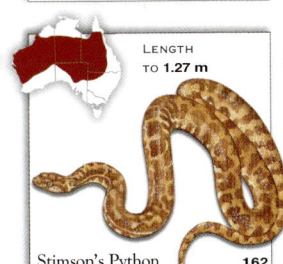

LENGTH
TO **1.27 m**

Stimson's Python   **162**

LENGTH
TO **8.5 m**

Scrub Python   **167**

LENGTH
TO **2 m**

Green Tree Snake   **172**

SNAKES

LENGTH
TO **1.5 m**

Slatey-grey Snake   **173**

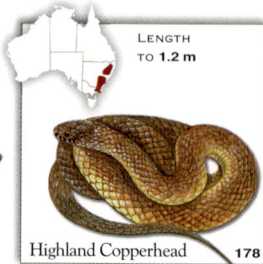

LENGTH
TO **1.2 m**

Highland Copperhead   **178**

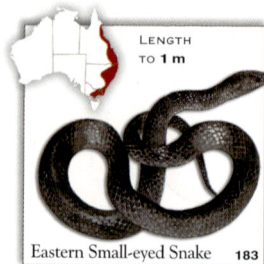

LENGTH
TO **1 m**

Eastern Small-eyed Snake   **183**

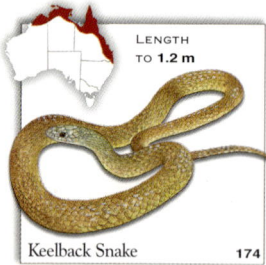

LENGTH
TO **1.2 m**

Keelback Snake   **174**

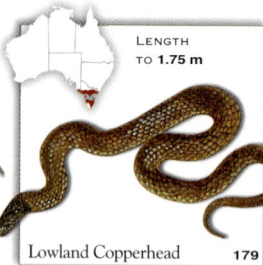

LENGTH
TO **1.75 m**

Lowland Copperhead   **179**

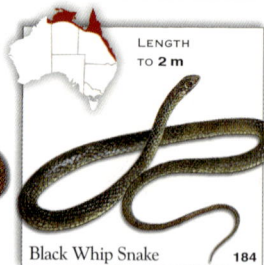

LENGTH
TO **2 m**

Black Whip Snake   **184**

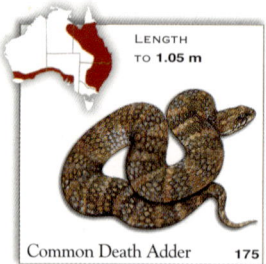

LENGTH
TO **1.05 m**

Common Death Adder   **175**

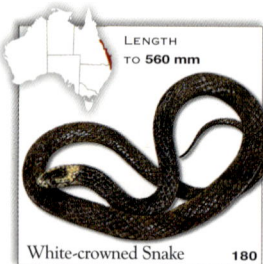

LENGTH
TO **560 mm**

White-crowned Snake   **180**

LENGTH
TO **1 m**

Yellow-faced Whip Snake   **185**

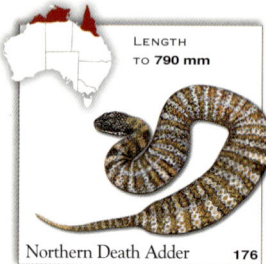

LENGTH
TO **790 mm**

Northern Death Adder   **176**

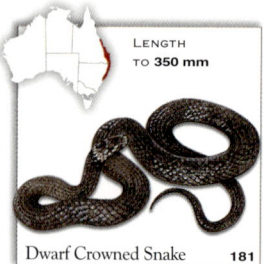

LENGTH
TO **350 mm**

Dwarf Crowned Snake   **181**

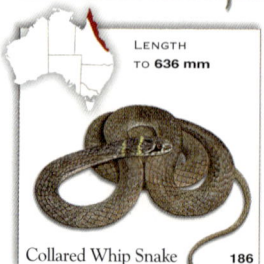

LENGTH
TO **636 mm**

Collared Whip Snake   **186**

LENGTH
TO **750 mm**

Desert Death Adder   **177**

LENGTH
TO **850 mm**

Golden-crowned Snake   **182**

LENGTH
TO **600 mm**

De Vis' Banded Snake   **187**

SNAKES

LENGTH
TO **500 mm**

White-lipped Snake 188

LENGTH
TO **2.4 m**

Black Tiger Snake 193

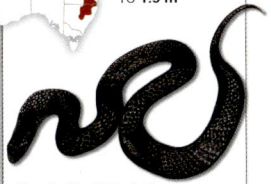

LENGTH
TO **1.9 m**

Blue-bellied Black Snake 198

LENGTH
TO **400 mm**

Red-naped Snake 189

LENGTH
TO **2 m**

Eastern Tiger Snake 194

LENGTH
TO **2.54 m**

Red-bellied Black Snake 199

LENGTH
TO **700 mm**

Orange-naped Snake 190

LENGTH
TO **2.7 m**

Inland Taipan 195

LENGTH
TO **2.13 m**

Dugite 200

LENGTH
TO **914 mm**

Black-bellied Swamp Snake 191

LENGTH
TO **2.9 m**

Coastal Taipan 196

LENGTH
TO **600 mm**

Ringed Brown Snake 201

LENGTH
TO **1 m**

Pale-headed Snake 192

LENGTH
TO **3.3 m**

King Brown Snake 197

LENGTH
TO **1.6 m**

Western Brown Snake 202

SNAKES

LENGTH
TO **2.2 m**

Eastern Brown Snake **203**

LENGTH
TO **750 mm**

Bandy-bandy **207**

LENGTH
TO **330 mm**

Desert Banded Snake **204**

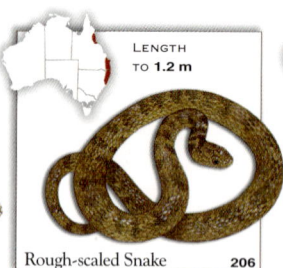

LENGTH
TO **1.2 m**

Rough-scaled Snake **206**

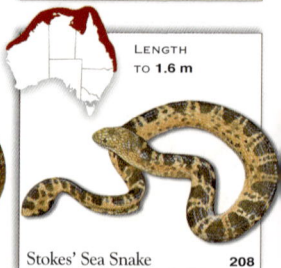

LENGTH
TO **1.6 m**

Stokes' Sea Snake **208**

LENGTH
TO **880 mm**

Curl Snake **205**

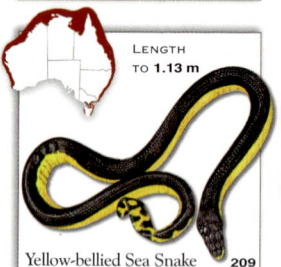

LENGTH
TO **1.13 m**

Yellow-bellied Sea Snake **209**

CROCODILES

LENGTH
TO **3.2 m**

Freshwater Crocodile **210**

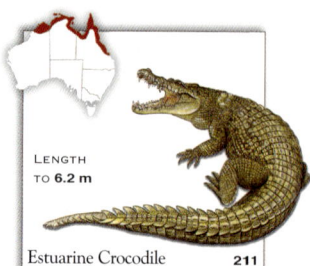

LENGTH
TO **6.2 m**

Estuarine Crocodile **211**

TURTLES

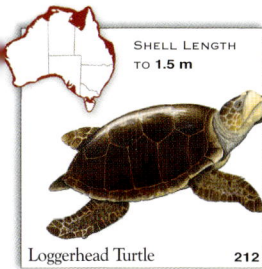

SHELL LENGTH
TO **1.5 m**

Loggerhead Turtle 212

SHELL LENGTH
TO **300 mm**

Eastern
Long-necked Turtle 216

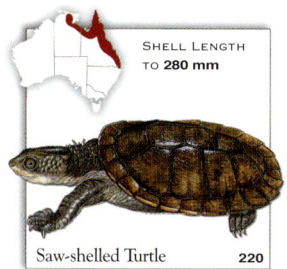

SHELL LENGTH
TO **280 mm**

Saw-shelled Turtle 220

SHELL LENGTH
TO **1.5 m**

Green Turtle 213

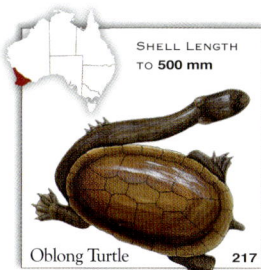

SHELL LENGTH
TO **500 mm**

Oblong Turtle 217

SHELL LENGTH
TO **340 mm**

Krefft's River Turtle 221

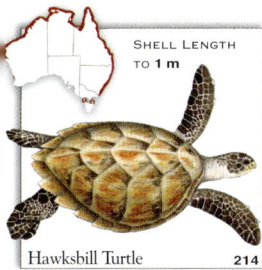

SHELL LENGTH
TO **1 m**

Hawksbill Turtle 214

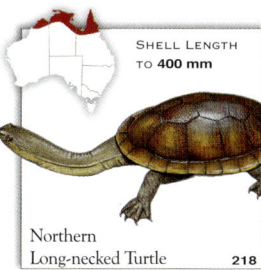

SHELL LENGTH
TO **400 mm**

Northern
Long-necked Turtle 218

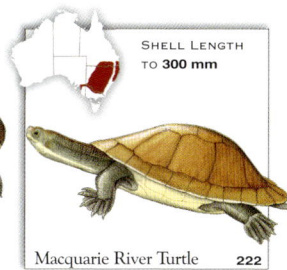

SHELL LENGTH
TO **300 mm**

Macquarie River Turtle 222

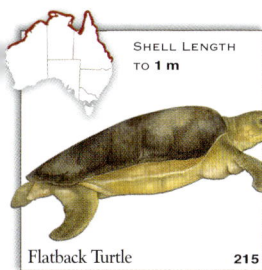

SHELL LENGTH
TO **1 m**

Flatback Turtle 215

SHELL LENGTH
TO **450 mm**

Northern Snapping Turtle 219

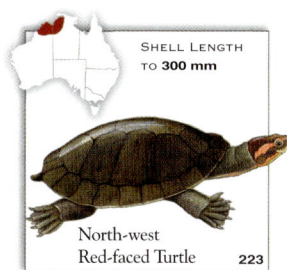

SHELL LENGTH
TO **300 mm**

North-west
Red-faced Turtle 223

FAMILY **MYOBATRACHIDAE**   SPECIES *Adelotus brevis*

TUSKED FROG

A medium-sized, flattened frog, with horizontal pupils, recognised by its striking black and white underside and bright orange-red markings on the groin, thighs and feet. It has long, narrow fingers and toes, and short legs with a trace of webbing between the toes. The back is rough and warty, black to a drab olive green or brown and mottled with irregular black spots forming a butterfly shape between the eyes. It has a dusky grey throat flecked with white. Males have a wide, flat head almost half the body size; a pair of small, sharp, fang-like tusks to 5 mm long, at the front of the lower jaw (only visible when the mouth is open); and a black belly with white spots. Females are smaller than males, with smaller tusks and a marbled black and white belly. **Behaviour** This very cryptic, ground-dwelling frog, hides by day under logs and rocks, beneath ground litter, in cracks and crevices, or in a burrow close to puddles, creeks and streams. In the breeding season males congregate around ponds or slow-flowing streams and call from concealed sites behind rocks, logs or vegetation. They sometimes create a nest at the water's edge by pushing leaves and other organic material around with their large head and tusks to form a cup-like depression, and call while sitting on the nest. When their numbers are high, males may fight over suitable nesting sites, using their tooth-like tusks to bite other males around the throat. They can often be heard calling by day and night, producing a short 'tok' or 'cluck', repeated at about 10 second intervals. **Development** They mate in late spring and summer, and females lay 140–485 eggs (about 1.8 mm across) in a floating foam mass. The egg mass is usually hidden among vegetation, or on leaves in the nest constructed by the male. The male usually stays next to or beneath the eggs until they hatch up to 10 days later. The tadpoles probably swim through the nest to the pond, where they grow to about 30 mm long before metamorphosing into 10 mm long froglets at 50–114 days.

Diet Insects, snails and other invertebrates.

Habitat Rainforests, wet sclerophyll forests and wet farm paddocks along the coast and ranges. They are usually found under logs, stones or leaf litter near puddles, ponds and streams.

Threats Habitat destruction, water pollution, fungal diseases and predation of eggs and tadpoles by introduced fish species.

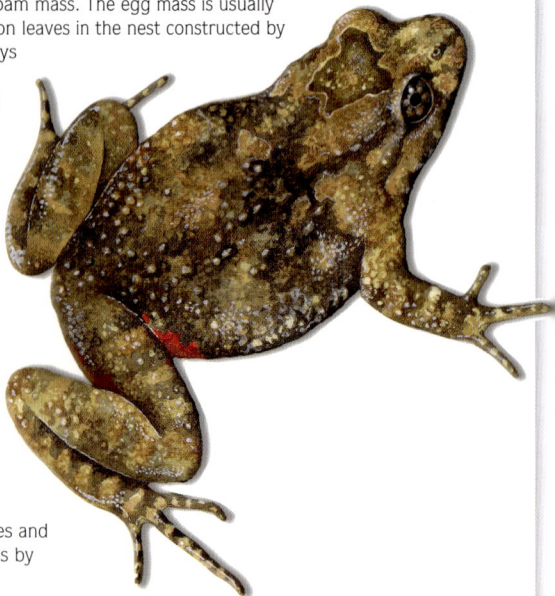

LENGTH **To 45 mm**   STATUS **Endangered in NSW. Vulnerable in Qld.**

FAMILY **MYOBATRACHIDAE** SPECIES *Crinia signifera*

EASTERN FROGLET

A small frog with a horizontal pupil and no webbing on the fingers and toes, although the toes are fringed. The colour and pattern are highly variable, even within a breeding group. It is commonly light grey or pale to dark brown above, blotched with black patches. Some individuals have black sides or a V-shaped bar between the eyes. The skin of the back is smooth or has irregular small warts, and there may be brown, black or sandy-coloured raised skin folds along the back, forming irregular longitudinal stripes. The belly is granular, whitish, mottled with prominent dark grey flecks, varying to marbled black and white. There is a small white spot at the base of each arm and males have a dark throat and chest. **Behaviour** This ground-dwelling frog shelters beneath rocks, logs and vegetation, and in soil cracks around swamps, ponds, slow-flowing streams and other wet sites. It emerges to breed after rain, and males may be heard calling day or night at any time of the year. Males call while floating among vegetation in shallow water, from around the edges of streams, producing a single 'creek' repeated about every second. A chorus of males is usually heard during and after rain, often along walking tracks and railway cuttings. **Development** They breed year round in most areas with a peak from July to September. Females lay 65–275 eggs (about 1.5 mm diameter), singly or in small, dark, jelly-like clumps of up to 30 eggs. They are usually attached to submerged grass stalks, twigs, leaves and other vegetation, in temporary or permanent waterways. The tadpoles vary in colour and grow to about 30 mm long before metamorphosing into froglets at about 49 days. Most frogs breed in their second year. **Diet** Beetles, ants and other small invertebrates. **Habitat** Most habitats including coastal wetlands, wet and dry forests, grasslands, disturbed areas, alpine meadows and urban areas. It is always associated with water.

| FAMILY **MYOBATRACHIDAE** | SPECIES *Heleioporus australiacus* |

GIANT BURROWING FROG

A large, robust, frog with a warty body, sometimes mistaken for the cane toad. It has a broad, rounded head with a distinct eardrum and large bulging eyes with vertical pupils. The limbs are short, the fingers have no webbing, and the toes are stubby with a trace of fleshy webbing and a large, spade-like pad on the sole of each hind foot, used for digging. Breeding males have swollen forelimbs with one large and several small black conical spines on the first fingers, and smaller spines on second and third fingers, possibly used in territorial defence. It is dark chocolate brown to bluish-black or slate-grey above, sometimes with white or creamish-yellow spots on the sides. A short yellow stripe runs along a ridge below the eye and eardrum. Adult males have small warts on the back and sides, each bearing a tiny black spine. The belly is granular, white or bluish-white. The throat is brownish, and there is often a yellow patch on the inner elbow. **Behaviour** This ground-dwelling frog searches for food over a wide area, often far from water, and is usually seen on warm, wet nights. Daylight hours are nearly always spent below ground in unformed burrows, under logs or other vegetation. In the breeding season they migrate to breeding sites where males move into crayfish burrows by the water's edge, or dig a burrow up to 560 mm long with a chamber at the end. They use their powerful hind legs to burrow into the ground, spiralling backwards, even into clay. Males call during and after heavy rain at night, from the partially flooded burrow entrance, or while hiding in dense waterside vegetation, or even when buried, producing an owl-like 'oo-oo-oo' sound. **Development** They breed mainly in late summer and autumn after heavy rain. Mating takes place in ephemeral pools and soaks, or rarely in permanent ponds or streams. Females lay up to 1250 large eggs in a foamy mass in the burrow chamber or in a foam raft concealed among vegetation at the water's edge. The eggs hatch 4–10 days later, or when the burrow floods and flushes the tadpoles into a creek. The tadpoles are slow-moving, dark and plump with a short tail and reach 80 mm long before metamorphosing into froglets at 3–11 months. **Diet** Insects, spiders, centipedes, crayfish and larvae. Tadpoles graze on algae at the bottom of deep, clear pools. **Habitat** Open, mainly dry forests and heaths of the coast and ranges, often on ridges, to about 920 m. It is found on sandstone shelves, around sandy creeks and hanging swamps when breeding, otherwise it is not dependent on wet habitats. **Threats** Habitat disturbance and destruction, pollution, fungal diseases, predation by foxes and cats.

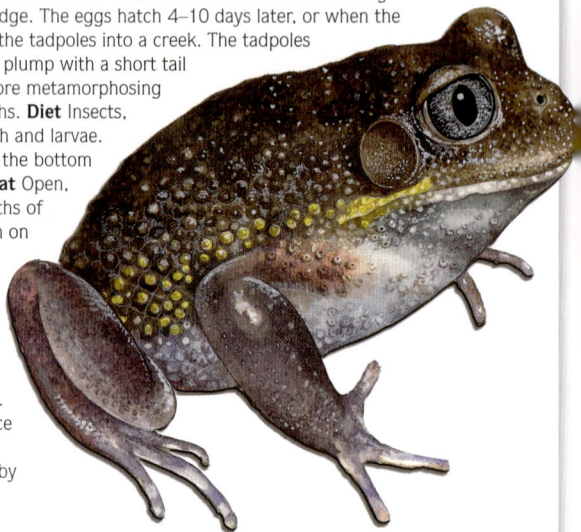

| LENGTH **To 90 mm** | STATUS **Vulnerable.** |

FAMILY **MYOBATRACHIDAE** SPECIES *Lechriodus fletcheri*

FLETCHER'S FROG
SANDPAPER FROG

A medium-sized frog of the wet forests. It has a flattened body, a horizontal pupil and a trace of webbing on the toes and fingers. The tips of the digits have very small discs and there is a round, hard pad on the soles of the hind feet. The back varies in colour from light grey to rich reddish-brown or dark olive-brown, usually with a few darker blotches and a dark bar between the eyes. A lighter ridge of skin runs from the nostril, above the eye, across the eardrum to the shoulder, with a thin darker stripe below it. The limbs have faint dark grey-brown bars, while the sides of the thighs are silvery grey. The skin is smooth or slightly granular above, although breeding males have small spines scattered over the back, giving them a sandpapery texture. The belly is smooth and whitish. The lower lip is brown.

Behaviour This ground-dwelling frog is well camouflaged and blends in with the texture and colour of its surroundings. It is usually found among leaf litter on the forest floor and in tree hollows. It is active from late September to late summer if there is enough rain, and is mostly seen during summer rains when they are breeding. Males call while floating on the water or on the ground near water during and after rain, around permanent and temporary pools, streams and small puddles, some as small as a cup, including water-filled hollows among tree roots. The call is a purring 'gar-r-r-up' lasting about one second. **Development** They breed in spring and summer and mate at the water's edge. They are prolific breeders, and females produce many clutches of up to 300 eggs (1.7 mm diameter) in a year. The eggs are laid in the water, and the female froths the mucus with her forefeet to make a foamy egg mass that floats on the surface. The tadpoles are small and translucent or dark brown with gold specks. They develop rapidly before the pool dries out, and metamorphose into young froglets at about 5 weeks, when they are some 35 mm long.

Diet Adults eat insects and other small animals. The tadpoles become cannibalistic if they eat all the plant and animal material in the pond, and only the largest survive. **Habitat** Wet sclerophyll forests and rainforests, around streams and ponds.
Threats Habitat modification and destruction. Urbanisation and tourism.

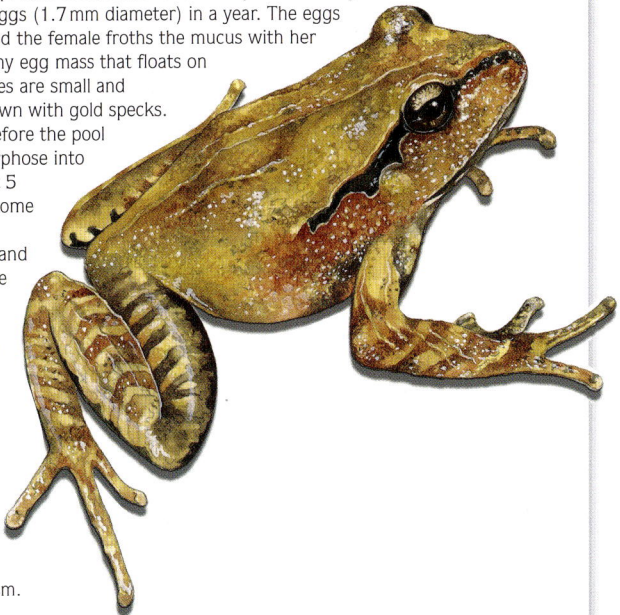

LENGTH **To 55 mm** STATUS **Rare in Qld.**

FROGS

| FAMILY **MYOBATRACHIDAE** | SPECIES *Limnodynastes convexiusculus* |

MARBLED FROG

A relatively large, stout, tropical frog with a horizontal pupil and no visible webbing on the hands and feet. The first and second fingers are the same length, and there are no discs or pads on the tips of the fingers and toes. It is grey, brown or dark olive above, marbled with dark blotches, and the skin has many low, flat glands, giving it a lumpy feel. There is a dark patch below the eye and irregular dark bands across the legs. The belly is smooth and white with dark mottling on the throat. The skin secretes a glue-like substance when irritated that causes snakes and other predators to release them. **Behaviour** This ground-dwelling frog has an unusual humpback posture. In the long dry season it retreats to the few permanent swamps or lives underground, digging a burrow or taking over an existing hole. Just after the first major rains of the northern wet season males congregate around low-lying swamps, often in large numbers, and call from hiding places in the grass, ground litter or the flooded burrows of crabs and other animals. The call is a high-pitched squeaky 'pung' repeated every second or so. **Development** They breed from October to March, and females lay their eggs in small, floating foam nests under dense vegetation or in a small puddle. Breeding females have flattened fingertips used to froth the mucus around the egg mass to create a foam raft surrounding the eggs. The black tadpoles are fully aquatic and are washed into nearby ponds and swamps by the rain. They attain a length of 70 mm before metamorphosing into froglets. **Diet** Small ground insects and spiders. **Habitat** Lowland coastal scrubs and savannah woodlands, preferring swamps and long grass. Also in urban areas.

| LENGTH **To 55 mm** | STATUS **Low risk.** |

30

FAMILY **MYOBATRACHIDAE**   SPECIES *Limnodynastes dumerilii*

EASTERN BANJO OR POBBLEBONK FROG

A moderate to large frog, stout and squat with a broad, rounded head, a horizontal pupil, short muscular hind legs and a large oval gland on the calf. The toes may be a quarter webbed or have only a trace of webbing. The first finger is equal to or slightly shorter than second, and there is a large, hard, spade-like pad on the sole of each hind foot. There are 5 subspecies of this frog with a great variation in colour, size and mating call. The back is usually dark brown, varying to grey, olive-green or black, with irregular dark markings, and sometimes with a pale stripe down the mid back. A distinct pale cream to yellow stripe runs beneath the eye and through the eardrum to the shoulder. The sides often have a bronze to bluish sheen mottled with black. There is a conspicuous pale ridge at the corner of the mouth. The skin above is smooth or slightly rough, with low, rounded warts. The belly is smooth and whitish, sometimes mottled with dark grey. The groin is white or yellow, and some individuals have a bright yellow throat. **Behaviour** This ground-dwelling frog is active most of the year and is often found on wet nights foraging in open areas, on roads and around dams and swamps. It hides during the day beneath a submerged bank or in a burrow dug at dawn into the bank of a pond, and emerges at dusk after sufficient rain. It uses the spade-like pads on its hind feet to dig into the mud, descending backwards into the ground. Males can be heard calling at any time of year. Their banjo-like 'plonk' or 'kuk-kuk' is repeated at intervals, and is often answered by others to create a chorus at slightly different pitches. They call while half submerged, resting on vegetation in the water, or from an air pocket in the flooded burrow. The gland on the calf produces a toxin that may deter some predators. **Development** They breed in spring, summer and autumn, and females lay up to 4000 eggs (1.7 mm diameter) in a floating, frothy mass, up to 1.8 m across, usually concealed among aquatic vegetation. The female beats air into the egg mass with her forefeet to create the froth. The tadpoles hatch a few days later and are dark grey with mottled tail fins. They develop slowly over the summer and metamorphose into young frogs as the water begins to dry up in February, when they are about 65 mm long. In colder areas the tadpoles take up to 15 months to develop. **Diet** Insects and other invertebrates. **Habitat** Most habitats including heaths, sclerophyll forests, rainforests, woodlands and grasslands along the coast, ranges and adjacent slopes, close to permanent waterholes, swamps, dams and ditches. They tend to avoid flowing creeks.

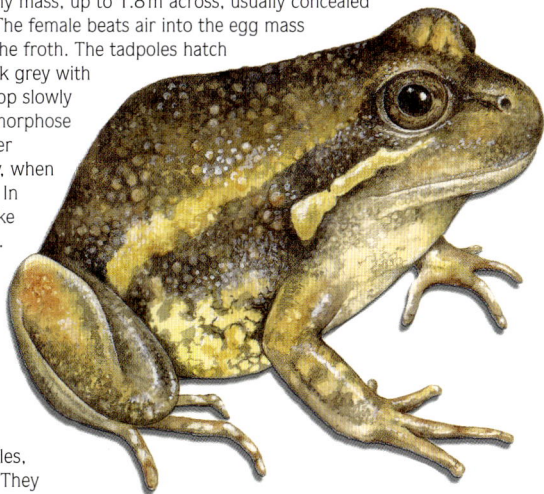

LENGTH **To 70 mm**   STATUS **Low risk.**

FAMILY **MYOBATRACHIDAE**   SPECIES *Platypectrum ornatum*

ORNATE BURROWING FROG

A medium-sized, stout, rather globular frog, with a short, rounded snout and a broad head with large eyes and lozenge-shaped pupils. The tips of the fingers and toes have no discs or pads. The toes are about one quarter webbed and there is a spade-shaped pad on the inside edge of the hind foot. The skin is smooth above, dotted with small lumps, sometimes tipped with red or cream. The back is light to dark yellowish-brown or dark grey with irregular darker blotches. There is usually a large butterfly- or U-shaped pale brown patch on the back of the head, and inland individuals often have a cream stripe down the middle of the back. The belly is smooth and white and the legs are short and stout with dark brown bands. **Behaviour** This ground-dwelling frog is commonly found foraging around river flood plains and roadways on wet spring and summer nights. It hides during the day in a burrow, using the spade-like pads on its hind feet to dig into the ground, shuffling its hind limbs and gradually pushing itself backwards into soft, sandy soil. In dry conditions it stays below ground for long periods, appearing suddenly after heavy rain to breed. Unlike some of the other burrowing frogs it does not form a cocoon around itself, and remains semi-alert in its burrow. Males congregate around permanent and semi-permanent ponds or puddles after heavy rain, often in large numbers, and call while floating freely on the surface with outstretched limbs, making a short, rapidly repeated 'unk' call. **Development** They breed in the wet season in the north and in spring and summer in the south. Females lay up to 1650 eggs (about 1.7 mm diameter) in a floating foam mass which disintegrates as the embryos hatch. The tadpoles develop quickly, and in shallow ponds may begin their metamorphosis into young frogs when only 3 weeks old. This rapid development allows them to use temporary pools that dry out quickly in the sun.

Diet Mainly ants, beetles, spiders, termites and other arthropods.

Habitat Dry coastal and inland areas including woodlands, grasslands and savannah, often along dry sandy watercourses some distance from permanent water. Also in areas subject to seasonal flooding.

LENGTH **To 45 mm**   STATUS **Low risk.**

BROWN-STRIPED FROG
STRIPED MARSH FROG

One of the most common frogs on the east coast, it is medium-sized with muscular hind limbs, a moderately pointed snout and a flattened body. The eye has a golden upper iris, a dark brown lower iris and a lozenge-shaped to round pupil. The toes have a trace of webbing and there are no discs or pads on the tips of the fingers and toes. The skin is smooth and slippery all over, light brown or grey-brown above with a pattern of irregular light and dark stripes and spots. There is often a distinct pale yellow to cream or orange stripe running along the middle of the back from the tip of the snout to the vent. A broad dark band runs from the snout through the eye and eardrum to the base of the forelimb with a white or yellowish raised stripe below it. The limbs are scattered with dark spots and irregular bands. The belly is white, often flecked with brown. **Behaviour** This ground-dwelling frog is a powerful jumper, and is associated with slow-moving streams, dams, swamps and marshes. It can be found year round, although it is less active in the cold winter months, and often takes up residence under debris on river flats and in garden ponds. Males can be heard throughout the year, particularly after heavy rain when they call relentlessly for many hours, often in chorus with others, their loud 'tok' or 'whuck' calls sound like a table tennis rally. They call while floating on the water almost completely submerged, from below the floating egg mass, or from well-concealed sites among waterside vegetation. **Development** They breed year round in the north and in the warmer months in southern areas. Females lay about 1000 eggs (1.5 mm diameter) in a foamy mass on the surface, usually among vegetation, in swamps, marshes, dams and ponds. Breeding females have a flap of skin on the second finger of each hand used to froth the water during egg-laying to create the foam raft. Tadpoles are black and reach 65 mm before metamorphosing into young frogs between 12 days and 11 months old. **Diet** Insects, other arthropods and small frogs. **Habitat** Most still freshwater habitats along the coast and ranges, including rainforests, wet and dry forests, swamps, dams, flooded grasslands and garden ponds.

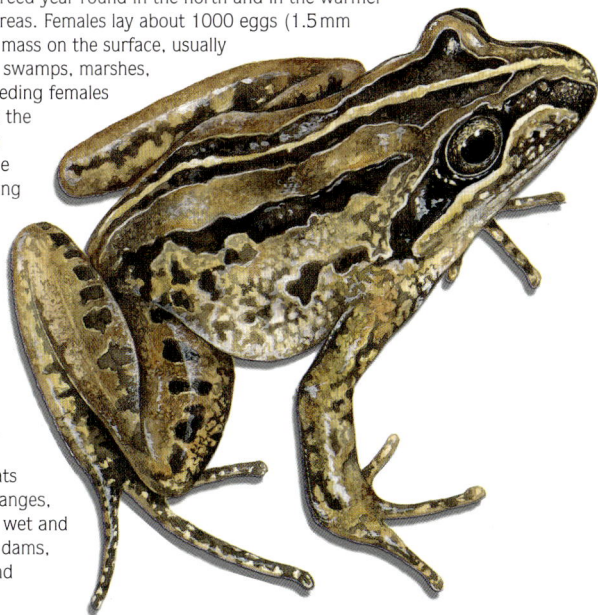

LENGTH **To 70 mm** STATUS **Endangered in Tasmania.**

| FAMILY **MYOBATRACHIDAE** | SPECIES *Limnodynastes tasmaniensis* |

SPOTTED GRASS FROG
SPOTTED MARSH FROG

A fairly small frog with a round to lozenge-shaped horizontal pupil and a golden iris. There are no discs or pads on the tips of the fingers and toes, and the second finger is longer than the first. The toes are fringed and slightly webbed at the base. This frog is distinguished by its distinctive pattern of greenish blotches and a white, pale yellow or russet stripe along the middle of the back from the snout to the vent. A dark, raised band bordered below by white or yellow runs from the snout to the eye and across the eardrum to the shoulder. The back and limbs have a background colour of light brown to a rich olive-green. The belly is white, and adult males have a darker throat. The skin is smooth above and below with a few low warts on the back. There is great variation in coloration, and in the far south of its range the frogs have narrow, bright red stripes down the mid-back.

Behaviour This ground-dwelling frog is active at night and is commonly found around dams and roadside ditches. It rests during the day beneath logs, rocks or debris near the edge of temporary and permanent ponds, swamps and creeks. After sufficient rain, at any time of the year, males congregate around water in low-lying areas and call while floating in the water in exposed or concealed sites, or from grassy vegetation around the edge of the water. Males make a repeated staccato 'kuk-kuk-kuk-kuk' in northern areas, or a single sharp 'click' or 'plock' call in southern parts. **Development** The main breeding season is from August to March, and females lay up to 1500 eggs (1 mm diameter) in a foamy mass 50–80 mm across floating among vegetation. She produces foam by forcing bubbles into the jelly-like substance surrounding the eggs with her paddle-like hands as she lays them. Females in the southeast of SA seem unable to synchronise their hand movements well enough to force many bubbles into the egg mass, and their eggs are laid in a small, foam-free clump. The tadpoles vary from light to dark golden brown to translucent gold, and metamorphose into young frogs at 3–5 months. **Diet** Insects, other arthropods and even small snakes.

Habitat Occupies a wide variety of habitats, from the dry interior to the wet coast, but avoids sandstone areas. It is often found in marshy areas, flooded grasslands, streams and ponds with grassy edges, and is quick to colonise new ponds and wetlands in disturbed areas.

| LENGTH **To 50 mm** | STATUS **Low risk.** |

FAMILY **MYOBATRACHIDAE** SPECIES *Limnodynastes terraereginae*

NORTHERN BANJO FROG
SCARLET-SIDED POBBLEBONK FROG

A large, stout frog with a lozenge-shaped horizontal pupil and a large, glandular swelling on the lower hind limbs. There are no discs or pads on the tips of the fingers and toes, and only a trace of webbing at the base of the toes. The inside of the soles of the hind feet have a large, shovel-shaped pad. The skin is smooth or slightly granular above, and smooth below. The back is grey or brown with scattered darker flecks and blotches, sometimes with a stripe down the mid-back. A raised orange or cream stripe runs from below the eye to the arm, with a dark band above, running from the snout, through the eye to the arm. The flanks usually have a reddish-orange stripe, and the upper arms have a reddish splash. The belly is white or pale yellow, and there are red or scarlet markings in the groin and on the backs of the thighs. **Behaviour** This ground-dwelling frog spends most of the year in an underground cavity at the base of a burrow where it shelters from the heat and desiccating effects of the dry climate it lives in. The large pads on the soles of the hind feet are used to scrape soil out from under the body when digging, so that it descends rear-first into the soil. It is usually seen on the surface after enough rain has fallen to penetrate through the soil to its underground cavity. In the wet season males call from concealed holes in the bank of a pond or swamp, while partially submerged in the still water. The call is a short, repetitive, high-pitched 'gonk' or 'donk', like someone plucking a banjo. **Development** They breed after sufficient rainfall between October and May. Females lay a foamy egg mass on the surface of still water, created by frothing the jelly around the eggs with their hands. The tadpoles are dark brown to black and metamorphose into young frogs at about 10 weeks. **Diet** Insects and other arthropods. **Habitat** Variety of habitats along edges of permanent streams, ponds, dams and swamps, surrounded by dense vegetation.

Duckworth

LENGTH **To 75 mm** STATUS **Low risk.**

FAMILY **MYOBATRACHIDAE**   SPECIES *Mixophyes balbus*

STUTTERING FROG
SOUTHERN BARRED FROG

A large, stout frog with a distinct eardrum but no discs
or pads on the tips of the fingers and toes. The toes are
three-quarters webbed. The eyes have a pale blue crescent
above and are darker below with an elliptical pupil. The body
is light brown to yellow-grey above with irregular darker
blotches and thin, dark, often poorly defined stripes across
the limbs. Unlike the great barred frog there are no conspicuous dark spots on
the sides. A narrow dark band runs from the tip of the snout (interrupted by
the nostril) across the top of the eye, and ends just behind the eardrum. The
belly is smooth, white to pale yellow. The backs of the thighs are speckled with
black. Females are larger than males. **Behaviour** This ground-dwelling frog is
active on wet nights and is usually seen close to permanent running water. After
heavy rain in late spring and early summer males gather in small groups and call
from leaf litter along the banks of creeks, making fast stuttering 'op-op-op' or
'kra-a-a-ah krook krook' calls lasting 1–2 seconds. They become territorial and
will grapple with each other at this time. Females construct a number of nests
(depressions scraped in gravel or leaf litter) along a river bank or in shallow
water, where mating takes place. **Development** They breed in December and
females lay some 250–550 sticky eggs (2.8 mm diameter) in cohesive clumps
in their nests or pasted directly onto bedrocks in the stream. If the nest is
among leaf litter on the bank of
a fast-flowing rainforest creek
the eggs are washed into the
water during the next heavy
rainfall. The eggs hatch
in the water 7–14 days
later. The tadpoles
metamorphose into
young frogs when
they are about 65 mm
long and about one
year old. **Diet** Insects,
arthropods and any
animal it can swallow,
including smaller
frogs and snakes.
Habitat Temperate
and sub-tropical rainforests
and wet sclerophyll forests along the coast and
nearby ranges to 900 m, particularly around
small permanent streams. **Threats** Habitat
destruction and disturbance to catchments
by logging, grazing stock and
pollution; predation by introduced
fish; fungal disease.

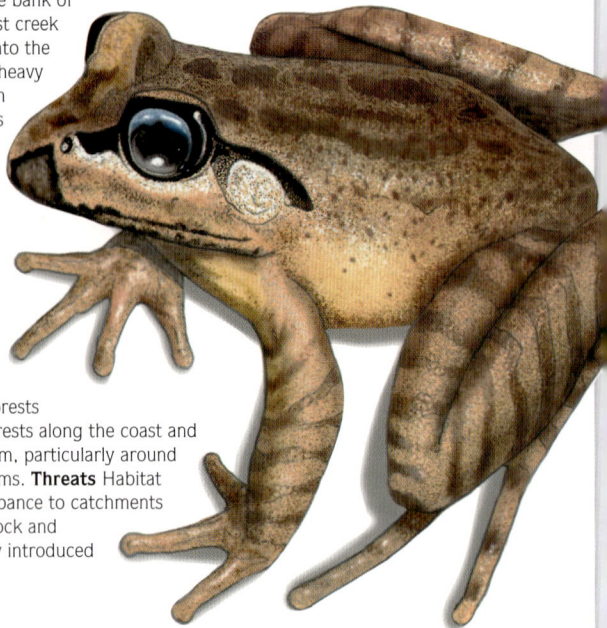

LENGTH **To 80 mm**   STATUS **Endangered in NSW. Critically endangered in Vic.**

GREAT BARRED FROG

A large, stout frog of the moist, coastal forests. It has a broad head, wide mouth and powerful hind limbs with dark cross bands that form triangular patterns on the inside. It has a vertical pupil, and the upper part of the iris is reddish-brown to black. The tips of the fingers and toes lack discs or pads. The toes are three-quarters webbed. The colour above varies from dark grey to light brown or bronze, mottled with darker markings and a series of dark blotches and spots along the sides. An irregular, sometimes broken, dark stripe runs along the back, starting between the eyes. The upper lip is pale cream, and a conspicuous dark stripe runs from the snout (broken by the nostril) above the eye to end just beyond the eardrum. The underside is smooth, white or pale yellow, with dark flecks on the throat and chin. **Behaviour** This ground-dwelling frog is well camouflaged and difficult to spot on the forest floor. It shelters by day beneath leaf litter or in a small burrow excavated in loose soil. It is usually found close to permanent running water and jumps away very quickly if disturbed or threatened. Males call in the warmer months from leaf litter along a creek bank. The call is a loud, deep, harsh, 'wark-wark-wark-wark' followed by a short, guttural, 'ruckle-ruckle'. The first part of the call is thought to attract females, while the second part may deter other males from its territory. **Development** They breed while floating in the water. The female lays a small number of eggs surrounded by a sticky jelly, and these are fertilised by the male as they emerge. She deposits the eggs on the webbing of her hind foot and flicks them onto the overhanging bank where they stick. The couple repeat this many times until 1000–2000 eggs have been deposited on the bank, safely out of the way of aquatic predators. When the eggs hatch the tadpoles fall or are washed into the water. They are dark brown to grey-brown and use their mouths to attach to rocks and other surfaces in flowing water. They grow to 65 mm long before metamorphosing into young frogs. **Diet** Insects, other arthropods and smaller frogs.

Habitat Rainforests and wet sclerophyll forests along the coast and nearby ranges.

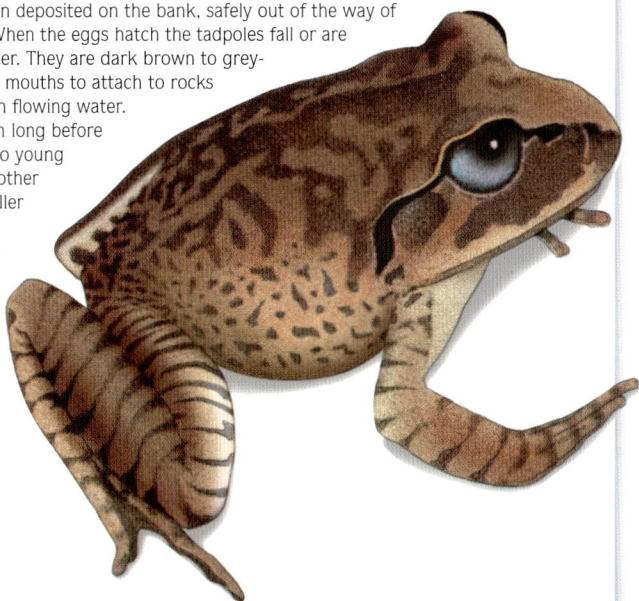

| FAMILY **MYOBATRACHIDAE** | SPECIES *Myobatrachus gouldii* |

TURTLE FROG

Perhaps the strangest-looking frog in the world, this species resembles a baby turtle without a shell. It has a plump, globular, flattened body, a small head and short, muscular legs with spade-like hands and plump digits. The skin is covered with small rounded lumps, and is dull pinkish-grey to brown above, with yellow flecks on the legs, and whitish below flecked with dark brown. The pupil is horizontal to circular and the toes are unwebbed. **Behaviour** This burrowing frog is seldom seen, and spends most of its time underground in a deep burrow dug into sandy soil. This is one of only two Australian frogs to burrow head first. The tip of the nose is protected by a small pad, while its spade-like hands and powerful front legs do the digging. It is found around termite nests and sometimes under logs and other surface debris, or on the surface after rain. Large numbers of males can often be heard calling together to attract females in late spring. They call from the surface or when partly buried in the soil with only their heads exposed. The call is a single creaky 'ba-a-a-ar-k' repeated every 1–2 seconds. Male and female pairs burrow into the sand where they mate and lay their eggs. **Development** Females lay 23–38 large eggs (about 5–7 mm across) in a burrow dug into moist sand, up to 1.2 m deep. The tadpoles develop underground entirely in the egg capsule, feeding on the rich yolk. The young frogs emerge and dig their way to the surface during the first autumn rains. **Diet** Feeds entirely on termites, eating more than 400 at one time. **Habitat** Dense scrubs with sandy soils and sand hills in semi-arid and arid areas, often in banksia woodlands, often associated with termite colonies or buried under logs and rocks.

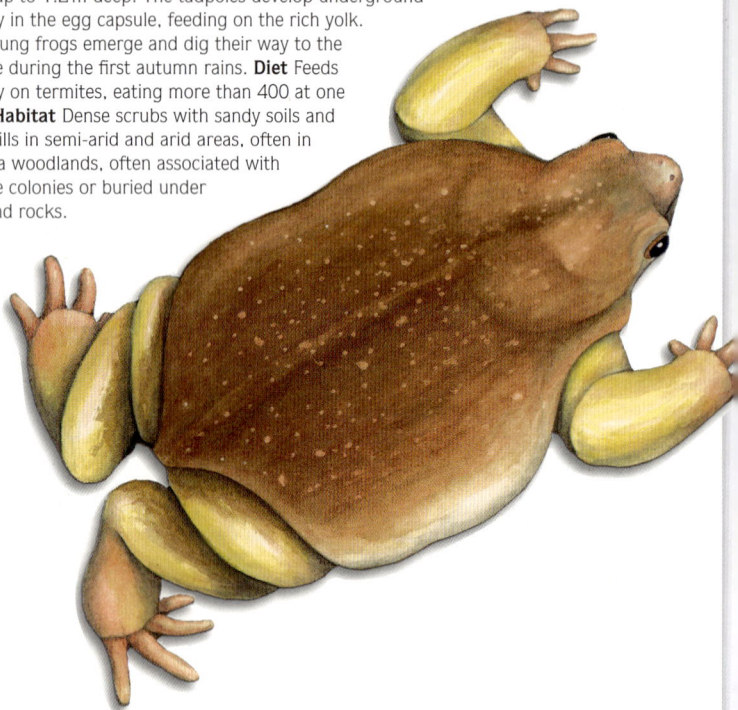

| LENGTH **To 60 mm** | STATUS **Low risk.** |

FAMILY **MYOBATRACHIDAE**   SPECIES *Neobatrachus sudelli*

PAINTED BURROWING FROG
COMMON SPADEFOOT TOAD

A small, plump, stout frog with stubby limbs. It has well-divided, fully webbed toes with large, hard, black pads on the outer edge of the soles of the hind feet for digging. It has a vertical pupil and is rough and warty above. Breeding males develop small spines on the back. The back is grey, yellow or reddish-brown, scattered with dark brown or olive spots and blotches. Most individuals have a narrow, pale, white, cream or yellow stripe along the spine. The belly is smooth and white. The skin around the groin is loose and baggy, and extends to the knee. **Behaviour** This ground-dwelling frog of the semi-arid regions is active at night after rain and spends the day in a burrow or buried under deep surface litter where it is cool and moist. It burrows into soft sand and mud using its feet as shovels, descending backwards, rotating its body as it digs. To avoid desiccation in the hot dry season it remains buried for long periods and becomes inactive, shedding a number of layers of skin to form a waterproof, translucent cocoon around itself. It emerges to feed and breed after sufficient rain has fallen to penetrate its burrow, and is usually only seen on the surface at night after summer rains. Males call after heavy rainfall while floating in temporary pools, swamps, clay-pans, ponds, dams and small creeks, and occasionally while still underground. The male has a large mouth cavity but no vocal sacs to resonate and amplify its call, which is a short trilled, clicking 'cr-aw-aw-aw-awk'. **Development** They mate in marshes, dams and ditches in autumn, late winter and early spring after substantial rainfall. Females lay a clutch of around 600–1000 eggs (1.5–1.9 mm diameter) on the water surface in long chains coated with jelly. These chains become caught up in water plants when present. The eggs hatch 3–10 days later into silvery-grey to golden brown tadpoles that grow to about 77 mm long before metamorphosing into young frogs at 4–7 months. **Diet** Grasshoppers, crickets, moths, flies and other arthropods. **Habitat** Dry regions in woodlands, shrublands, mallee, open grasslands and disturbed sites.

LENGTH **To 40 mm**   STATUS **Low risk.**

FAMILY **MYOBATRACHIDAE**   SPECIES *Notaden bennettii*

CRUCIFIX TOAD
HOLY CROSS FROG

A plump, distinctive frog with a horizontal pupil and an almost spherical body with short limbs, a short snout and small mouth. The toes are slightly webbed with a large, hard, pale, shovel-shaped pad on the inside edge of the soles of the hind feet. The first finger is shorter than the second. The back is warty, olive, yellow or green, with a cross-shaped pattern of red, black, yellow and white spots. The sides are usually adorned with scattered yellow, black and white spots. The belly is smooth and white. Juveniles are bright yellow or emerald green. **Behaviour** This ground-dwelling frog of the semi-arid regions is usually only seen and heard after heavy summer rain when the males call while floating in temporary pools. The call is a series of long, owl-like 'whooos', rising slightly in pitch. At the slightest disturbance it sinks quickly out of sight. It is a poor jumper, but can run with remarkable speed. It squats on ant and termite trails and catches them with indiscernible flicks of its tongue. When threatened it inflates its internal vocal sacs with air, deterring predators by its size and distinctive colours. It can also secrete a sticky, yellow, poisonous fluid from glands on its back. The fluid dries into a strong, elastic mass, entangling and irritating an attacker. It spends most of the year underground in a deep burrow to escape the hot, dry conditions, digging into moist soil with its hind feet, descending rear first in a circular motion. **Development** They breed in shallow temporary pools after summer rains, and the females lay about 500 eggs in a floating oval mass, usually surrounded by a protective band of vegetation. They hatch 2–3 days later into grey-brown or dark brown tadpoles. They develop quickly and grow to 40 mm long before metamorphosing into young frogs at 28–30 days. **Diet** Small black ants, termites and other small invertebrates.

Habitat Open woodlands, mallee, savannah and black-soil flood plains on the western plains and slopes. Mostly underground, only emerging after heavy rain.

LENGTH **To 55 mm**   STATUS **Low risk.**

FAMILY **MYOBATRACHIDAE**   SPECIES *Paracrinia haswelli*

HASWELL'S FROGLET
RED-GROINED FROGLET

A relatively small frog with a horizontal pupil, fringed toes without webbing and occasionally with very slightly enlarged fingertips. Its most conspicuous features are the bright orange to red patches in the armpits, groin and backs of the thighs. It is light grey-brown to brown above with irregular darker patches or flecks. A paler stripe often runs from between the eyes down the centre of the back, and a dark band runs from the nostrils through the eye and eardrum to the shoulder. The skin is smooth above and slightly granular on the belly, which is light to dark grey with paler patches.

Behaviour This ground-dwelling frog is active at night and usually rests by day in rushes and reeds growing at the edges of waterways, and is sometimes found under stones in rocky creek beds. Males call from August to March, usually after rain, while floating concealed among grasses and sedges or from the cover of litter and vegetation on the banks of waterways and ditches. The call is a short, explosive 'ank' or 'ark', repeated at about 10 second intervals, and sounds like ducks quacking in the distance.

Development They breed on the margins of streams, swamps or dams from spring to early summer. Eggs are laid in loose clusters and attached to the stems of plants below the water's surface. The tadpoles vary from translucent yellow to yellowish-brown or grey and are usually found in warm shallow water hiding among water plants. They metamorphose into froglets about 13 mm long at about 100 days. **Diet** Insects and other arthropods. **Habitat** Wet and dry sclerophyll forests, woodlands, shrublands and coastal heaths, among vegetation bordering creeks, coastal swamps, ponds and dams.

LENGTH **To 35 mm**   STATUS **Low risk.**

FAMILY **MYOBATRACHIDAE**   SPECIES *Pseudophryne australis*

RED-CROWNED TOADLET

A small, rather stout frog with short limbs and characteristic black and white marbling on the belly. The toes are unwebbed and the pupil is horizontal. The skin is smooth or with a few low warts on the back, and is dark brown to black, often with a red tinge. There is a bright red, orange or rarely yellow triangular patch on the crown of the head, a similar-coloured or white patch on the upper forearms, and a red or orange stripe along the lower back. **Behaviour** This secretive, ground-dwelling frog hides during the dry summer months deep in cracks and burrows in the ground, and emerges in early autumn after good rainfall to feed and breed. Its short limbs make hopping difficult, and it tends to walk with short hops. They congregate in colonies of up to 50 individuals around seepages and rock crevices below sandstone cliffs and ridges, also in gutters beside tracks and fire trails, hiding under rocks, logs and other surface debris. The call is a short, soft, squelch-like 'eeeek-eek' repeated several times in quick succession at infrequent intervals. The calls can be heard year-round.
Development They breed in autumn and females lay a clutch of about 24 eggs in moist crevices, under a log or moist leaf litter alongside a small pool or stream, often scraping out a simple nest on the ground. The male stays close to the developing eggs to defend his breeding site. The tadpoles are small and dark to light grey and develop within the egg capsule, feeding on the yolk, until the hind legs appear. Development stops at this stage until water flows through the nest after heavy rainfall, allowing the tadpoles to break out of the egg capsule and swim into small pools. They spend most of their time on the bottom of the pool among the leaf litter, and metamorphose into young frogs when they are about 8 mm long. Fecundity is low with few tadpoles reaching metamorphosis. **Diet** Ants and other small arthropods.
Habitat Open forests, woodlands and heaths. Confined almost totally to Hawkesbury sandstone formations within a radius of about 160 km around Sydney. **Threats** Urban development, habitat degradation and fragmentation, pollution, removal of bushrock, fungal disease.

LENGTH **To 30 mm**   STATUS **Vulnerable.**

Family **Myobatrachidae**   Species *Pseudophryne coriacea*

Red-backed Toadlet
Red-backed Broodfrog

A small, stout frog with a horizontal pupil and short limbs, without webbing on the toes or pads on the fingertips. Like other members of the genus it has black and white marbling on the belly and can be distinguished by its striking rich brown to bright-red back, contrasting with a broad black band along the sides of the head and body. The back has dark flecks and there is usually a white patch at the base of the arms, and sometimes a pale patch on each side. The skin above is smooth or velvety, sometimes with a few low warts, while the belly is smooth or slightly granular to touch. Females are larger than males. **Behaviour** This ground-dwelling frog has short legs and is only able to crawl or make short hops. It usually hides among leaf litter, logs and other debris. Males tend to stay within a small area, while females roam further afield. In the breeding season males call from the banks of creeks or swamps, hiding in a shallow burrow dug into the bank or sometimes in the open. They also call from the cover of grass, beneath moist leaf litter or under rotting logs. The call is a short, grating 'ark' or a short, sharp 'squelch'. They can be induced to call by making a loud sharp shout. **Development** They breed after rain in spring, summer and autumn. Females lay a mass of 40–110 large eggs (up to 5 mm across) in a moist burrow close to the water's edge. The eggs hatch when the burrow is flooded by heavy rainfall and the small, grey-brown tadpoles develop in about 2 months in temporary pools. Males often stay with the eggs and call from the nest to attract other females to mate with. They have a lifespan of 4–7 years. **Diet** Ants, termites, greenfly and other small arthropods. **Habitat** Rainforests, wet and dry sclerophyll forests, woodlands and occasionally cleared sites along the coast and ranges in low-lying marshy areas, inundated ditches, beside creeks and temporary pools.

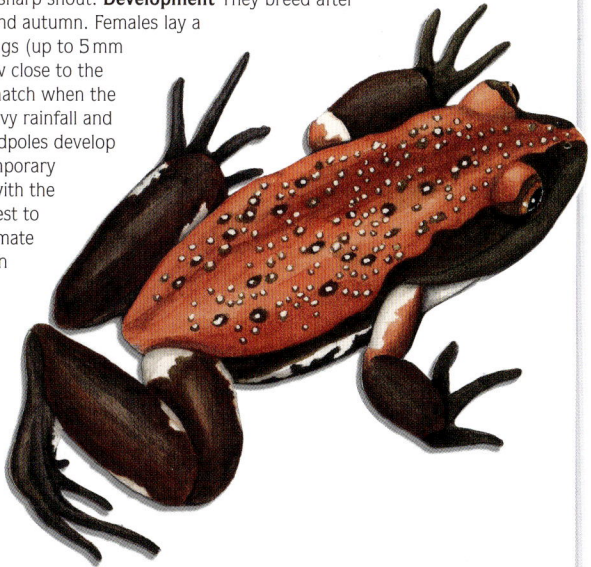

Length **To 35 mm**   Status **Low risk.**

FAMILY **MYOBATRACHIDAE**   SPECIES *Pseudophryne corroboree*

SOUTHERN CORROBOREE FROG

A small, stout frog of the cold, wet alpine regions, easily identified by its striking black and yellow markings resembling the body painting of Aborigines attending a corroboree. It has a horizontal pupil, short limbs with unwebbed toes and no pads on the fingertips. The belly is marbled black and white or black and yellow. The skin is slightly granular above with low warts, and smooth below. Northern individuals are much less yellow, and sometimes appear lime-coloured or green. **Behaviour** This ground-dwelling frog has difficulty swimming and hopping, and generally crawls around on all-fours. They are most active after rain from October to November, and can be found beneath logs and leaf litter in alpine forests. They move down to pools around sphagnum bogs and seepage areas to breed during January and February, and are active mainly at night. Males dig shallow burrows 50–100 mm apart in thick moss beside shallow pools, and stay in their burrow, hidden from predators, during the summer breeding season. They call to attract females and to keep other males out of their territory, making short, harsh, squelching calls. Each male will attract up to ten females to his burrow sequentially and may dig a new burrow if his first is filled with eggs. The colder months of the year are spent hibernating beneath the snow. **Development** They breed in late summer and females lay 12–50 relatively large eggs (about 3 mm diameter) in the male's burrow. The eggs hatch in 4–5 weeks and the tadpoles develop in the jelly-like mass, which prevents them drying out. They feed on the yolk until they are in an advanced state within the egg coat and then cease development until the burrow is flooded by melting snow or rain, some months later. The tadpoles then swim to the nearest pool and continue their growth, metamorphosing into young frogs in December or January, almost 12 months after fertilisation. Young frogs move to the alpine heaths and become sexually mature at 3–4 years. They have a lifespan of up to 9 years.

Diet Adults are generally ant-eaters, juveniles eat any small arthropods.
Habitat Montane and sub-alpine regions from 1300–1760 m in heaths, snow gum woodlands, wet sclerophyll forests with a dense understorey, wet tussock grasslands and sphagnum bogs. **Threats** Habitat degradation, tourist development, drought, fungal disease.

LENGTH **To 30 mm**   STATUS **Critically endangered.**

FAMILY **MYOBATRACHIDAE** SPECIES *Uperoleia laevigata*

SMOOTH TOADLET
EASTERN GUNGAN

A small frog with short legs and fringed toes without any webbing. It has an elliptical pupil and enlarged glands on the sides of the head. The glands produce a poison that deters predators. The skin is granular or rough and warty above, brown or olive-brown, with light and dark brown spots and blotches, and distinct reddish patches in the groin and behind the knees. There is usually a prominent pale triangular patch on top of the head. The belly is smooth, grey to creamish and speckled with dark purplish-brown. **Behaviour** This ground-dwelling frog is active in the warmer months of the year. It forages on the ground at night, and hides by day beneath leaf litter or in a burrow dug into soft soil. Males call after sufficient rainfall in spring, summer and autumn, congregating in a small area in an open site or among low vegetation at the edges of flooded grasslands, dams or ditches, or while floating in the water. The call is a grunting or harsh grating 'aaaaaahk' lasting about 1 second and repeated every 2–3 seconds. Over a period of 3–4 nights, females move slowly through the congregation of calling males before deciding which one to mate with. **Development** They mate in spring and summer and females lay their eggs (about 1.3 mm diameter) singly or in small clumps attached to submerged vegetation at the bottom of still, shallow water. The tadpoles are about 22 mm long, mottled gold and black or golden brown.
Diet Insects and other arthropods.
Habitat Drier forests and woodlands in grassy sites and disturbed areas that become flooded after rain.

LENGTH **To 32 mm** STATUS **Low risk.**

45

| FAMILY **HYLIDAE** | SPECIES *Cyclorana alboguttata* |

STRIPED BURROWING FROG
GREEN STRIPE FROG

A large, relatively slender frog with a horizontal pupil and expanded, grooved discs on its toes and fingertips. The toes are half-webbed, and the inside edges of the hind feet have a hard, spade-shaped pad. It is brown, olive or green above, scattered with darker blotches, warts and ridges. A green or pale yellow stripe usually runs down the back, and a dark streak runs from the snout, through the eye and eardrum, expanding into a broad, broken band on the sides. The upper edge of the streak has a distinct fold of skin. The backs of the thighs are very dark with white or yellow spots. The belly is finely granular and white. The throat and chest are smooth with brown flecks. Females are larger than males. **Behaviour** This ground-dwelling frog is usually only seen around temporary pools and water filled clay-pans. It is active both day and night, and is able to climb vertical surfaces as smooth as glass, gripping with the grooved discs on its toes and fingertips. To avoid dehydration in dry conditions it burrows into the ground using the spade-shaped pads on its hind feet, descending backwards, digging itself into the soil at an angle. Water loss is further reduced by adopting a water-conserving posture with its head lowered and limbs folded under its body. In this position it forms a translucent waterproof cocoon around itself by shedding several layers of skin, leaving only its nostrils exposed, and slows down its metabolic rate. When rain penetrates its underground chamber the frog eats its cocoon and digs its way out. Males congregate on the grassy edges of temporary ponds and ditches after heavy summer rain, and make a loud, rapid, quacking call to attract females. **Development** They breed in the wet season after good rainfall, and females lay a sheet of eggs over submerged rocks or on the bottom of a pond. The eggs hatch into large tadpoles in less than 36 hours. **Diet** Insects and other small arthropods. **Habitat** Coastal and inland woodlands, grassy and cleared areas, flood plains, dams and waterholes.

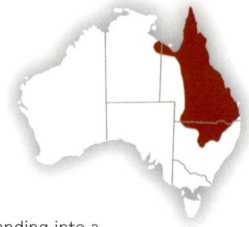

| LENGTH **To 70 mm** | STATUS **Low risk.** |

FAMILY **HYLIDAE** SPECIES *Cyclorana australis*

GIANT FROG
NORTHERN WATER-HOLDING FROG

Australia's largest ground-dwelling frog, it has a horizontal pupil, powerful limbs, a bulbous body and a relatively flat head. The toes are webbed at the base and have a hard, shovel-shaped pad on the sole. The first finger opposes the others for grip, but the fingertips have no discs or pads. Its colour varies from grey to sandy brown, dull brown or occasionally dull pink above, sometimes with bright green patches, while juveniles are often emerald green. A dark stripe runs from the snout through the eye and eardrum to the shoulder. The skin above is smooth with a few scattered warts or folds, and a distinctive fold runs from above the shoulder towards the thigh. The underside is white and finely granular, with some brown flecks on the throat. The groin often has a pale blue-green tinge, and the backs of the thighs have a brown, cream or dark blue pattern. **Behaviour** This burrowing frog shelters from the dry, desert heat in a cavity at the base of a deep burrow with its bladder full of water. (In emergencies they can be dug from dry ponds and almost pure water squeezed from them.) To avoid dehydration it encloses itself in a waterproof cocoon made from repeated sheddings of its skin, leaving only its nostrils uncovered for respiration, and slows down its metabolism. It uses the shovel-shaped pads on its hind feet to dig, pushing itself backwards into the soil, and only emerges after the first heavy rains of the wet season. It is most active at night, but may be seen basking in the sun at the edge of shallow, temporary pools, or crossing roads after tropical storms. It spends most of its time sitting still, waiting for insects to come within range of its sticky tongue. Males call from the edge of semi-permanent waterholes and dams, hidden below an overhanging rock or bank, making a loud, deep, 'wok' or honking call. **Development** They breed from late November to February and females lay 100–1000 eggs (about 1.5 mm diameter) in clumps in small pools and swamps, loosely attached to plants. The tadpoles are large, cream or dark olive grey and metamorphose into young frogs before the pond dries out, when less than 30 days old. Young frogs sit beside the pond and eat other emerging frogs. **Diet** Mainly grasshoppers, beetles, ants and termites, but will eat any animal it can catch and swallow, including its own species. **Habitat** Coastal flood plains, grasslands, woodlands and monsoon forests.

LENGTH **To 100 mm** STATUS **Low risk.**

SHORT-FOOTED FROG
SUPERB COLLARED FROG

A small, stout frog of the arid areas of eastern Qld. It has
a horizontal pupil, toes that are one quarter webbed, and
a shovel-shaped pad on the inside edge of the soles of the
hind feet. The skin above is smooth or finely granular with a
few low warts. It is rich brown, marbled with lighter patches
on the back, sides and limbs, with a narrow stripe along the
mid back from the snout to the vent. A broad dark streak runs from
the nostril to the eye and continues across the eardrum to the shoulder.
The belly is finely granular and whitish. **Behaviour** This ground-dwelling
frog is usually seen near clay-pans after rain. To survive dry conditions
it buries itself deep underground close to temporary pools, and can live
without water in a chamber at the base of its burrow for at least a year.
It digs with the shovel-shaped pads on its hind feet, entering the ground
backwards. Once buried it sheds a number of layers of skin to create a
water-resistant cocoon around itself, leaving only the nostrils exposed
for respiration, and slows down its metabolism. When water from heavy
summer rains penetrates its underground chamber the frog eats the
cocoon wrapping for extra energy, and digs itself out. Thousands of frogs
congregate around temporary waterholes where they feed and mate,
producing long, loud 'craaaark' calls that can be heard from several
hundred metres. When the rains cease they burrow into the ground
again. **Development** They breed in the wet season and females lay long
strings of eggs in temporary ponds. The eggs hatch into tadpoles and
develop very rapidly before the waterhole dries up. The tadpoles are quite
large, translucent gold or grey with darker mottling, and
metamorphose into young frogs at
about 25 days. **Diet** Mostly insects
and other arthropods.
Habitat Mainly savannah
woodlands and open
grasslands in drier
areas along the
coast and inland.

FAMILY **HYLIDAE**   SPECIES *Cyclorana novaehollandiae*

North-eastern Water-holding Frog
New Holland Frog

A large frog with a bull-shaped head and a horizontal pupil. It differs from *Cyclorana australis* by its wider, more rounded jaw, lack of pattern behind the thighs and a dark band running vertically from below the eye to the upper lip. The toes are webbed at the base and the feet have a hard, shovel-shaped pad on the inside edge of the sole. The fingertips have no discs or pads and the first finger opposes the others for grip. It is pale grey, brown or yellowish above, sometimes flecked or blotched with dark brown. A dark brown stripe runs from the snout through the eye to the eardrum. The skin is smooth above, with a few scattered warts and distinct folds running along each side. It is whitish below, the throat is flecked or lined with grey or brown, and the backs of the thighs are grey or bluish. Juveniles may be bright green or have green blotches. **Behaviour** This ground-dwelling frog survives drought and heat by burying itself in a chamber at the end of a deep burrow. It sits in a water-conserving posture with its head bent down and its limbs tucked under its body, wrapped in a waterproof, parchment-like cocoon, made by shedding several layers of skin, leaving only the nostrils exposed for respiration, and slows down its metabolic rate. It emerges during heavy summer rainfall when water penetrates the burrow chamber. Large numbers congregate in temporary waterholes and floodlands to breed, the males making repeated short, resonant 'honk' calls to attract females. To avoid predators they sit partly submerged with only the nostrils above water. Coastal individuals may be active year round if there is sufficient rainfall.

Development They breed in temporary waterholes in the northern wet season, or year round in wetter, coastal areas. Females lay clumps of up to 1000 eggs in shallow, still or slowly moving water. The eggs sink and hatch into dark brown to golden brown tadpoles soon after laying. The tadpoles develop in 4–6 weeks, reaching 100 mm long, metamorphosing into young frogs before the waterhole dries out. **Diet** Mainly insects, occasionally geckos and any other animal it can swallow.

Habitat Woodlands, seasonally wet scrublands, rainforests, savannahs and coastal flood plains.

LENGTH **To 100 mm**   STATUS **Low risk.**

FAMILY **HYLIDAE**   SPECIES *Cyclorana platycephala*

WATER-HOLDING FROG

A moderately large, stout frog of the arid interior, with a flattened head and relatively small eyes with horizontal pupils. The toes are fully webbed and the inside edge of each hind foot has a spade-like pad. The skin is smooth above with a few warts and smooth or granular below. It has an olive-grey, olive-green or grey back with pale green patches and scattered darker flecks. The belly is whitish and the throat of breeding males is flecked with dark brown. **Behaviour** This ground-dwelling frog spends most of its time below ground where it is able to live for years during prolonged droughts. They absorb large amounts of water through their skin and store it in their bladder and in pockets under the skin. In emergencies, desert Aborigines sometimes squeeze out the water to drink. They burrow into damp, clay soil with spade-like pads on their hind feet, descending slowly backwards into the ground, and survive in a spherical chamber with compacted walls up to 1 m underground. To prevent water loss they shed several layers of skin to create a translucent waterproof cocoon, and slow down their metabolic rate, breathing slowly through exposed nostrils. Water from flooding rain soaks into the chamber and awakens the frog which eats its cocoon and digs its way back to the surface. They feed on the ground and under water, catching aquatic prey with the hands, building up fat reserves as fast as possible before the water evaporates. Males congregate around pools after rain and strike up a loud chorus of 'woarp-woarp-woarp' calls to attract females. **Development** They breed after good rainfall and females lay a large mass of up to 500 eggs in temporary pools and still waterways. The tadpoles feed on organic debris at the bottom of the pool, and metamorphose into young frogs at about 30 days when they are about 25 mm long. **Diet** Mainly ants and termites supplemented by any other terrestrial or aquatic animal they can catch and swallow.
Habitat Black-soil plains, grasslands, temporary swamps, clay-pans and billabongs in arid and semi-arid inland areas.

LENGTH **To 72 mm**   STATUS **Low risk.**

FAMILY **HYLIDAE**   SPECIES *Litoria aurea*

GREEN AND GOLDEN BELL FROG

A large frog, distinguished from *Litoria raniformis* by its smooth or slightly granular skin. It has a horizontal pupil and small, distinct discs on its toes and fingers, slightly wider than the digits. The second finger is longer than the first. The toes are almost fully webbed and the feet have a small hard pad on the inside edge. It is dull olive-brown to bright green above with splashes and lines of gold or bronze. A cream or gold stripe with a black edge runs from the nostril through the eye and along the sides to the groin, with a skin fold above it. The backs of the thighs and groin are bright blue or turquoise. The belly skin is coarsely granular and white. Males are smaller than females. **Behaviour** This ground-dwelling, semi-aquatic frog is active by day and night and spends most of its time in the water. It shelters among bulrushes and reeds growing in permanent lagoons and slow-moving waterways, and under debris in river flats. It has a voracious appetite and does not hesitate to eat its own species. It rarely strays far from its preferred feeding area except in the breeding season when it sometimes travels long distances searching for a mate. Males call while floating on the water after rain, mainly from August to January, making a series of groans followed by 'craw-awk, crawk, crok, crok' calls. **Development** They breed in summer and females lay a large floating cluster of up to 12,000 eggs among vegetation near the water's edge. The eggs hatch within 3 days into large, fast moving, yellow to golden tadpoles. They metamorphose into young frogs at about 6 weeks, although this can take as long as 11 months. **Diet** Cockroaches, water snails, yabbies and any other small animal it can catch and swallow, including other frogs. **Habitat** A variety of habitats around permanent streams, lagoons, swamps, dams and ponds with bulrushes and other emergent vegetation, often in disturbed sites including abandoned mines and quarries. **Threats** Habitat destruction, pollution, fungal disease and predation of its eggs and tadpoles by introduced fish.

LENGTH **To 85 mm**   STATUS **Endangered in NSW and the ACT. Vulnerable in Vic.**

FAMILY **HYLIDAE**   SPECIES *Litoria bicolor*

NORTHERN DWARF TREE FROG
NORTHERN SEDGE FROG

One of Australia's smallest tree frogs, it has a horizontal pupil and distinct discs on the fingers and toes. The fingers have a trace of webbing and the second finger is longer than the first. The toes are half to three-quarters webbed and fringed, and there is a small, hard pad on the inside edge of the hind feet. The skin above is smooth and satiny, green with a broad bronze band running along the centre of the back. A dark brown stripe runs along the side of the snout, through the eye, over the eardrum and along the flanks. A white or cream stripe runs along the upper lip to the shoulder. It is smooth below except for the belly which is granular and cream or yellowish. The groin and backs of the thighs are pale yellow. **Behaviour** Using the discs on its fingers and toes this frog is able to climb smooth, vertical surfaces. It is generally found on the stems and leaves of plants, often in hot, dry places, and unlike most other frogs it often basks for long periods in the sun on the exposed upper surfaces of broad-leaved shrubs. In the dry season it is frequently found in the upper leaf bases of pandans. It avoids drying out by keeping its limbs close to its body to reduce water loss by evaporation. In the wet season thousands gather among tall reeds and grasses where they hunt for food. Breeding males call at night from high up in the vegetation in or beside permanent and temporary ponds, lagoons and swamps. Their loud, high-pitched bleating 'ree-e-eck-pippip' call also attracts snakes and other predators.
Development They breed in the wet season, from December to March, and females lay small clumps of 10–20 eggs attached to submerged vegetation. The eggs are 1.9–2.9 mm diameter and hatch into pale brown tadpoles. The tadpoles are up to 55 mm long when they metamorphose into young frogs at about 11 weeks of age.
Diet Mainly insects and other arthropods.
Habitat Grasslands and open woodlands, near permanent semi-permanent waterways and marshes, also in suburban gardens along the coast and adjacent ranges.

LENGTH **To 30 mm**   STATUS **Low risk.**

FAMILY **HYLIDAE**   SPECIES *Litoria caerulea*

Green Tree Frog

A very large tree frog with a blunt, rounded snout, a thick glandular ridge above the eye, a horizontal pupil and a golden iris. It has large pads on the tips of the fingers and toes, and the feet have large hard pads on the inside and outside edges. The fingers are about one-third webbed, the toes are three-quarters webbed, and the second finger is longer than the first. The skin is smooth above and coarsely granular on the sides and belly. It is bright pale green to dark olive-green above, often spotted with white on the back or sides, and there is sometimes a stripe from the corner of the mouth to the shoulder. The colour can change in less than an hour to blend with the background. The belly is whitish to cream, and juveniles have a faint pale stripe above the eye. **Behaviour** This ground-dwelling and climbing species is often found around houses, using the large gripping pads on its fingers and toes to climb glass and other smooth, vertical surfaces. It is a powerful jumper, and often waits near outdoor lights where it pounces on insects. In the breeding season, when it is raining or the humidity is high, males call from rocks and small boulders beside waterways. They are also often heard calling from hollow tree limbs or drainpipes which amplify their rasping 'wark' or deep 'crawk' calls. **Development** They breed after heavy summer rain in the south and in the wet season in the north. Females lay 2000–3000 eggs, 1.1–1.4 mm diameter. The eggs are expelled with great force through the sperm cloud, and often end up half a metre away, landing as large clumps on the surface of pools or puddles, where they sink to the bottom. The tadpoles are dark green or brown and metamorphose into young frogs at about 38 days. They have a lifespan of up to 23 years. **Diet** Insects, spiders, mice and other small animals. **Habitat** One of the most widespread frogs, it lives in a variety of habitats around waterways from the coast to the dry interior, often in and around houses, water tanks, downpipes and toilets.

LENGTH **To 100 mm**   STATUS **Low risk.**

FAMILY **HYLIDAE**   SPECIES *Litoria chloris*

RED-EYED GREEN TREE FROG

A long, flattened tree frog with a horizontal pupil and a golden red iris. The second finger is longer than the first, and there are large pads on the tips of the fingers and toes. The fingers are three-quarters webbed and the toes are almost fully webbed. It is smooth to finely granular above, and bright leaf-green. The backs of the thighs are brown to purplish-red, often with paler iridescent purple patches on the upper edge. The upper arm, hands and feet are yellow, while the outermost finger and toe are usually green. The belly is granular to touch, white or bright lemon-yellow. **Behaviour** This climbing frog spends its life in the foliage, often high up in the canopy. It is very acrobatic and capable of catching insects in flight. It can also climb smooth, vertical surfaces, gripping with the pads on its fingers and toes. It is active at night, and only descends to the ground to breed, when large numbers gather in low-lying areas after heavy rain. The males call from shrubs and the lower branches of trees, eventually congregating in clearings around temporary ponds and swamps, producing an almost deafening chorus of long, moaning 'wark wark wark' calls followed by softer trills. **Development** They breed after heavy rain in spring and summer and females lay several small clusters of eggs attached to aquatic plants above or underneath the surface in still water. The tadpoles are light grey to dark brown and develop quickly before the pool dries out, metamorphosing into young frogs at about 41 days. **Diet** Mostly flying insects such as moths and flies. **Habitat** Wet sclerophyll forests, rainforests, heaths, wetlands, urban areas and regrowth areas along the coast and adjacent ranges, often on the flood plains of coastal rivers.

LENGTH **To 65 mm**   STATUS **Low risk.**

Blue Mountains Tree Frog

A medium-sized tree frog with bright red or orange markings on the backs of the thighs, in the groin, and the inside edge of the hind feet. It has a horizontal pupil and large discs on the fingers and toes. The fingers have little or no webbing and the second finger is longer than the first. The toes are nearly half webbed, and there is a large, hard, oval pad on the inside edge of the hind foot. The skin is smooth to finely granular above, sometimes scattered with warts, and light to medium brown with darker flecks. A dark brown stripe with a thin white upper border runs from the nostril, through the eye, along the sides almost to the groin. There are bright green areas along the sides of the head and body, and along the forelimbs and lower legs. The amount of green depends on the temperature and temperament of the frog, and may disappear completely. The belly is white and coarsely granular. **Behaviour** This ground-dwelling and climbing frog is active by night, and is usually only seen in the warmer months of the year. It shelters by day in rocky crevices, and prefers sites near large boulders and running water. In spring and summer adults congregate around ponds and the males call from rocks and vegetation close to the water's edge, making a soft chuckling or a harsh 'warrk' followed by short, rapid, 'cruk-cruk-cruk-cruk-cruk-cruk-cruk' calls. In winter they disperse and are often found well away from their breeding sites. **Development** They breed in permanent or semi-permanent pools, and females lay around 900 eggs (1.7 mm diameter) in the water. Once the eggs are laid, the female kicks the water with her hind feet to disperse them, and they spread out and sink. The small, brownish tadpoles are strong swimmers and metamorphose into young frogs at about 2 months. **Diet** Mainly insects and other arthropods. **Habitat** Wet and dry sclerophyll forests, rainforests, woodlands and heaths along the coast and ranges, usually around rocky creeks and stream beds with riparian vegetation.

DUCKWORTH

LENGTH **To 60 mm**   STATUS **Low risk.**

FAMILY **HYLIDAE**   SPECIES *Litoria dahlii*

DAHL'S AQUATIC FROG
NORTHERN WATER FROG

A large, slender, flattened frog with a pointed snout, horizontal pupils and long limbs. There are small discs on the tips of the fingers and toes, and the second finger is longer than the first. The fingers are unwebbed, whereas the toes are fully webbed. The skin is slimy to touch, finely granular above, green to olive-brown or olive-grey, often with darker markings, and usually with a pale green stripe running from the snout down the middle of the back. The backs of the thighs are mottled or spotted with white, and the belly is white. **Behaviour** This large, aggressive, ground-dwelling and semi aquatic frog is active by day, and is often seen floating in the water or basking in the sun on floating waterlily leaves. Large numbers congregate during the wet season around the edges of billabongs and swamps to breed, producing low, soft, short barking calls, sometimes while floating in the water. It hunts on the water surface, on land and also underwater. It is predated upon by snakes and lizards, and uses its sleek form and powerful legs to escape its enemies. It also exudes large amounts of mucus, making it very slippery and hard to grasp. The mucus is toxic to snakes and other frogs and has an acrid smell. In the dry season it hides in deep, moist soil cracks to escape the heat and avoid dehydration. **Development** Females lay large floating clumps of more than 1000 eggs in still water after suitable rainfall, mainly from January to March. The tadpoles are quite large and vary from dark to light grey-brown.
Diet Flies, ants, grasshoppers, other small invertebrates and small frogs, including members of its own species.
Habitat Savannah woodlands, often around the edges of permanent and semi-permanent water on the floodplains.

LENGTH **To 70 mm**   STATUS **Low risk.**

Family **Hylidae**   Species *Litoria dentata*

Bleating Tree Frog

A medium-sized frog with a long and somewhat flattened body, a rounded snout, a horizontal pupil and a dark red iris. The tips of the fingers and toes have moderate-sized discs, and there is a large, hard pad on the inside edge of the hind foot. The second finger is longer than the first. The fingers are almost one-third webbed and the toes are three-quarters webbed. The back is smooth and brown with a broad, paler band running along the flanks from the eyes to the lower back, and a dark band along the spine. The belly is granular and yellowish-white. The throat is smooth and lemon-yellow, becoming brown or black in breeding males. The sides, armpits and groin are often lemon-yellow, particularly in males. **Behaviour** This nocturnal, ground-dwelling and climbing frog hides by day under loose bark, in narrow crevices and hollows in trees, and beneath stones near waterways. It is sometimes transported in plant pots and building materials. Adults are usually only seen and heard after rain in spring and summer when large numbers of males congregate beside creeks, swamps and dams, and call from low shrubs and grass or while in the water. Their high-pitched bleating calls can become a deafening, pulsating chorus at the height of the breeding season when hundreds of males call together. **Development** They breed in shallow grassy and grass-edged swamps, usually after heavy spring and summer rains, from December to early March. Females lay small clumps of 230–3000 eggs attached to waterside vegetation. The eggs hatch quickly into dark brown tadpoles before the pool dries out. These are often seen swimming together near the surface. **Diet** Mainly insects and other arthropods. **Habitat** Paperbark swamps behind coastal sand hills, rainforests, wet sclerophyll forests, woodlands, heaths and urban areas on the coastal plains and adjacent ranges, usually around lagoons and ponds.

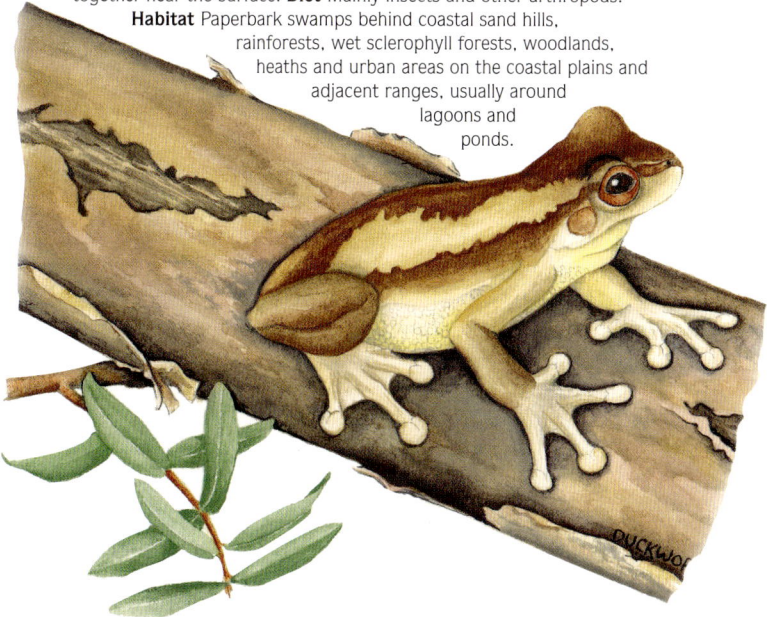

Length **To 45 mm**   Status **Low risk.**

FAMILY **HYLIDAE**   SPECIES *Litoria fallax*

EASTERN DWARF TREE FROG
EASTERN SEDGE FROG

A small, slender, elongated frog with a horizontal pupil
and moderate discs on the tips of the fingers and toes.
The fingers have only a trace of webbing, and the second
finger is longer than first. The toes are half to three-quarter
webbed and there is a small hard pad on the inside edge of
the hind foot. The skin is smooth and satiny above, green or
pale brown, changing according to the background colours of its resting site,
sometimes with scattered darker flecks. A distinct white stripe runs along the
upper lip below the eardrum to the shoulder. The belly is granular to touch
and cream or yellowish. The groin and backs of the thighs are bright orange
to yellowish. **Behaviour** This climbing frog is active both day and night, and
is often seen resting on plants growing in swamps, dams and lagoons. It also
travels well away from water, and can be found during the day sheltering
from the sun in the leaf bases of banana plants, pandans, pineapples and
bromeliads. These frogs are sometimes carried in produce from market gardens
to greengrocer shops. In southern areas they are active mostly in the warmer
months when males call from bulrushes, reeds and other vegetation fringing
still waterways, making repeated squeaky 'reeek-pip-reeek-pip-pip' calls, their
bright orange vocal throat sacs blown up like balloons. **Development** Mating
takes place in spring and summer in dams, swamps and ponds with plenty of
vegetation. Females lay small clumps of around 260 eggs in still water among
submerged plants and aquatic litter. The small tadpoles are brown to gold with
prominent, pigmented tails. As
they grow they change colour to
bright green and metamorphose
into young frogs at about 4
months when they are about
55 mm long. **Diet** Small insects and
other arthropods. **Habitat** Various
habitats around dams, swamps,
lagoons and slow-
moving creeks
along the coast
and adjacent
ranges.

LENGTH **To 25 mm**   STATUS **Low risk.**

FREYCINET'S ROCKET FROG

A medium-sized frog with a long, pointed snout. It has a horizontal pupil and small discs on the tips of the fingers and toes, only slightly wider than the digits. The second finger is the same length as the first. The fingers are unwebbed, while the toes are almost fully webbed. The skin is smooth above with low warts and rows of skin folds. It is mottled pale grey-brown to dark brown, with a distinct whitish stripe running below the eye and eardrum to the shoulder, and a broader, broken, black stripe from the snout through the eye and eardrum and along the sides. There is usually a pale triangular patch on top of the head with a darker blotch in the centre. The belly is whitish and granular, the throat is smooth. The backs of the thighs are brown with cream blotches. Females are larger than males.

Behaviour Aptly named, this largely terrestrial frog is very fast and agile, and quickly jumps out of the way if disturbed, leaping high and far. It is active at night and is usually found during the warmer spring and summer months close to swamps. Males congregate around ponds and waterways at this time of year after heavy rain, often in large numbers, and call from dense vegetation at the water's edge making sharp, loud, rapidly repeated, quacking or yapping sounds.

Development They mate in spring and summer after rain, the female depositing 200–1000 eggs in ephemeral ponds and swamps. The male kicks the eggs free as they emerge and scatters them over the water where they sink to the bottom of shallow water. The tadpoles develop quickly and can survive high water temperatures. They grow to about 35 mm long before metamorphosing into young frogs.

Diet Mainly insects and other arthropods. **Habitat** A wide variety of heath and forest habitats, particularly around temporary swamps along the coast and adjacent ranges. **Threats** Habitat destruction and fragmentation, weeds, land clearing, pollution of waterways.

LENGTH **To 42 mm** | STATUS **Vulnerable.**

FAMILY **HYLIDAE**   SPECIES *Litoria gracilenta*

DAINTY GREEN TREE FROG

A medium-sized tree frog with a horizontal pupil and a golden yellow or reddish iris. It has bright yellow feet with large flattened discs on the tips of the toes and fingers. The second finger is longer than the first. The fingers are about three-quarters webbed, while the toes are almost fully webbed. The skin is finely granular above, sometimes with a few small lumps on the sides and limbs, and coarsely granular below. The back is bright leaf-green to pea-green, although it can change its coloration to match the surrounding vegetation. The sides, upper arms, insides of the arms and legs, fingers and toes are bright lemon-yellow, while the backs of the thighs are maroon or reddish-brown, often with an iridescent blue sheen. A faint greenish-yellow stripe runs from the nose across the top of the eye and eardrum. The belly and throat are yellow to cream.

Behaviour This climbing frog is often found in boxes of bananas shipped around Australia. It is active at night and is usually seen on roads or among low vegetation after rain, and sometimes around house lights hunting for flying insects. It rests by day in dense streamside vegetation, and avoids detection by flattening itself against the foliage, holding its limbs tight against its body with its eyes closed and retracted, showing as much leaf-green colour as possible. This position also helps it avoid dehydration in dry conditions. After spring and summer rainfall males can be heard calling around waterways from floating vegetation, reeds, or from the foliage of trees or shrubs, making long 'waaa' or 'weee' calls. **Development** They breed in spring and summer and females lay 500–1000 eggs in a single layer or in clusters attached to vegetation in still water. The dark brown tadpoles metamorphose into young frogs when they are about 34mm long at 14 weeks. Froglets are golden brown and change to their adult colours within a few weeks. **Diet** Mainly insects and other arthropods.
Habitat Moist forests and woodlands along the coast, usually near creeks, swamps, lagoons and flooded areas. In winter it shelters in the crowns of trees well away from water.

LENGTH **To 45 mm**   STATUS **Low risk.**

GIANT OR WHITE LIPPED TREE FROG

The world's largest tree frog, it has a horizontal pupil and large discs on the tips of the fingers and toes. The second finger is longer than the first and the fingers are at least half-webbed. The toes are completely webbed with a large, hard pad on the inside edge of the hind foot. The skin is finely granular above, coarser on the sides and belly, and smooth on the throat. It is normally bright green above, but its colour changes to brown depending on the temperature and background. A conspicuous white stripe runs along the lower lip to the shoulder, and another stripe runs along the edge of the hind leg to the toes. It is whitish below. Females are larger than males. **Behaviour** A nocturnal climbing frog, it forages on wet and humid nights, and is often found around houses and sheds, in bathrooms and toilet cisterns where it climbs smooth, vertical surfaces using the gripping pads on its fingers and toes. After summer rainstorms males gather around swamps and ponds, often perching on trees, and set up a deafening chorus of harsh barking calls to attract females. When distressed they make a cat-like miaowing call. **Development** They breed in summer after rain and females lay some 430 eggs (about 1.9 mm diameter) in dumbbell shaped clumps in still water. The eggs hatch in 1–2 days into dark brown tadpoles with a cream stripe on each side of the head, body and tail. They metamorphose into young frogs at about 2 months and have a lifespan of up to 15 years. **Diet** Mainly insects and other arthropods.

Habitat Generally lives in low-lying areas in a wide variety of habitats including rainforests, heaths, swamps, mangroves, dry sclerophyll forests, seasonally dry monsoon woodlands, cultivated areas and around houses along the coast and adjacent areas.

FAMILY **HYLIDAE**   SPECIES *Litoria latopalmata*

BROAD-PALMED ROCKET FROG

A small frog with a pointed snout and horizontal pupils. It has very small discs on the tips of its fingers and toes. The fingers lack webbing while the toes are fully webbed except for the fifth toe which is less than half-webbed. The hind feet have a small hard pad on the inside edge and a minute pad on the outside edge of the sole. The skin is smooth above with a few scattered warts and granular below. There are large colour variations between populations and more than one species probably shares this name. It is generally pale fawn to dark brown above with variable soft, darker blotches. A dark streak runs from the nose through the eye and eardrum to the flank, broken by a pale blotch in front of the eye which curves down below the eye to the shoulder. Dark blotches mottle the sides and the groin is blotched with brown and yellow. The backs of the thighs are patterned with yellow and dark brown, while the belly is whitish. **Behaviour** This agile, ground-dwelling frog forages in the leaf litter at night, and avoids potential predators by making fast, high leaps if disturbed. It rests by day under the cover of rocks, logs and plant debris, and forages widely after dark over the forest or woodland floor, sometimes well away from water. Breeding males congregate around the edges of ponds and other watercourses including dams, flood plains and fast-flowing rivers on warm, wet nights in spring and summer. They call from the edge of a pond or nearby damp site, setting up a chorus of loud, sharp, repeated, quacking or yapping calls. **Development** They breed in spring and summer, and females lay loose clumps of up to 350 small eggs among vegetation in the shallow water of a pond. The eggs often sink to the bottom and hatch 3–6 days later into sandy gold tadpoles. They metamorphose into young frogs when they are about 35 mm long at 50–90 days. **Diet** Mainly insects and other arthropods. **Habitat** Open forests, woodlands, open country and river flood plains along the coast, ranges and western plains. Outside the breeding season it is often found well away from waterways.

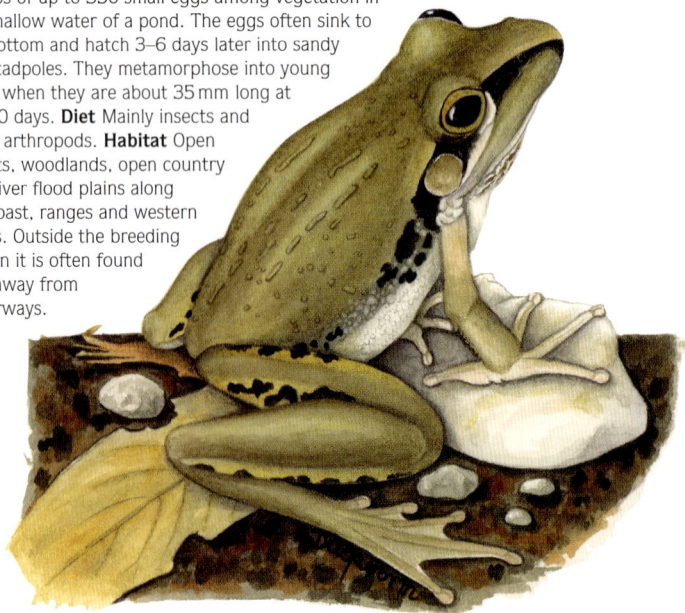

LENGTH **To 40 mm**   STATUS **Low risk.**

STONY CREEK FROG
LESUEUR'S FROG

A medium-sized frog, with a horizontal pupil and a streamlined body with long legs. It has very small discs on the tips of the fingers and toes. The fingers are unwebbed, while the toes have well-developed webbing. The skin above is smooth and varies from bright yellow in breeding males to pale fawn or dark brown, sometimes with a few scattered darker flecks and irregular black spots on the sides. A dark streak runs along the side of the head from the tip of snout, through the eye and over the eardrum to the shoulder, usually with a narrow pale border above. The skin is whitish below, granular on the belly and smooth on the throat. Populations north of Sydney have yellow and black groin markings, and those to the south have blue and black groin markings. Males are much smaller than females. **Behaviour** This frog is predominantly ground-dwelling, and rests during the day beneath rocks and in crevices by the water, or in the open, blending in with its surroundings. It is active at night and roams widely outside the breeding season, covering large distances, foraging for food among leaf litter and sometimes in low vegetation along the banks of waterways. It favours rocky creeks strewn with smooth boulders, and is often seen crossing roads on wet nights. It is very agile and escapes danger by making fast, powerful leaps. In the breeding season males find and defend territories close to the edge of creeks and ponds, choosing raised sites about 1–2 m apart. Rival males are kept away from the best calling sites by the agitated calling of resident males. They lack a vocal sac and produce a purring trill, repeated for 2–3 seconds, and begin calling some time before the females arrive. **Development** They breed from August to May, and mate underwater, lying on the bottom in the quiet reaches of a stream. The eggs (about 1.5 mm diameter) are laid in clumps of about 500 and stick to rocks or pebbles in flowing water. In northern Qld females make shallow nests in exposed sandy creek banks, about 250 mm across and 50 mm deep, and lay their eggs in the nest. **Diet** Insects, cockroaches, beetles, other arthropods and small vertebrates, including frogs.
Habitat Heaths, sclerophyll forests, woodlands, rainforests and urban areas, along the coast, nearby ranges and slopes, often around shallow rocky creeks, but may be found some distance from water.

| FAMILY **HYLIDAE** | SPECIES *Litoria nannotis* |

TORRENT TREE FROG
WATERFALL FROG

A robust frog with a broad, blunt, rounded snout, a speckled iris and a horizontal pupil. The tips of the fingers and toes have large pads. The second finger is longer than the first and the fingers are webbed at the base. The toes are almost fully webbed and there is a prominent hard pad on the inside edge of the hind foot. The skin above is satiny and finely granulated or scattered with warts. It is olive-green to brown or almost black, covered with darker flecks, and there is often a metallic bluish sheen on the flanks. The skin below is granular and whitish, often brown on the throat. The backs of the thighs are dark brown. Females are larger than males. **Behaviour** This cryptic, rock-dwelling frog, is active both day and night. It is found in and alongside fast-flowing rocky rainforest creeks and near or behind waterfalls. If disturbed it leaps into the torrent and is swept downstream to land on a rock where it remains motionless to prevent detection. It blends in perfectly with its surroundings with markings that closely resemble the moss and lichen-covered rocks in and around the water's edge. In the breeding season males develop large black spines on their thumbs and chest. These are used to grip the female while mating and prevent him being washed off her back. Males are also territorial and use complex visual displays to deter other males from their territory. The resident frog faces up to the intruder and waves its front legs in circular motions while calling loudly. They lack a vocal sac and utter short, harsh growls or repeated 'crawk-crawk-crawk' calls. **Development** They mate at any time of year in fast-flowing water and females stick their gelatinous egg mass beneath a submerged rock. The eggs are quite large and measure about 3 mm across. The tadpoles are sandy coloured with powerful tails and sucker-like mouths enabling them to grip onto rocks so they are not swept away. They grow to 51 mm before metamorphosing into froglets. **Diet** Mainly insects and other arthropods. **Habitat** Rainforests and wet sclerophyll forests of north eastern Qld, within or next to waterfalls and fast-flowing rocky creeks. **Threats** Fungal and viral diseases, feral pigs.

| LENGTH **To 59 mm** | STATUS **Endangered.** |

STRIPED ROCKET FROG

A medium-sized frog with a pointed snout and a horizontal pupil. It has long legs with small discs on the tips of the fingers and toes. The fingers are unwebbed, while the toes are about half-webbed. The skin above has many low, rounded warts and folds forming dark longitudinal raised ridges. It is variable in colour and pattern, usually pale to reddish-brown with irregular dark brown blotches or stripes and a white lip. A dark band runs from the snout through the eye and eardrum to the shoulder, with a cream bar in front of the eye. A pale stripe runs beneath the eye to the shoulder and sometimes to the snout. The skin is granular on the belly, smooth on the throat and chest and whitish. Males have brown speckles on the throat. **Behaviour** This fast, streamlined, ground-dwelling frog is active at night and forages in the leaf litter on the floor of open forests and around the edges of permanent swamps and creeks. On wet nights it is usually found around water in relatively open sites, where it uses its powerful leaping abilities to escape predators. When disturbed it jumps with remarkable speed, covering up to 4 m in one hop. Males call from the ground among vegetation on the banks of waterways, their 'yap yap' calls often creating a raucous chorus after heavy summer rain. **Development** In northern Australia they breed in temporary soaks and waterholes during the wet season, and in permanent swamps in the dry season. Southern populations breed in summer. Females lay clusters of 50–100 eggs (about 1 mm diameter) in a thin film of jelly which floats on the surface of still water or attaches to vegetation. The dark brown tadpoles take 1–5 months to metamorphose into young frogs. **Diet** A wide range of small, ground-dwelling insects and spiders.
Habitat Open forests, melaleuca wetlands, flooded grasslands, coastal swamps, streams and ponds, urban areas.

FAMILY **HYLIDAE**   SPECIES *Litoria peronii*

EMERALD SPOTTED OR PERON'S TREE FROG

A medium-sized frog with a horizontal pupil and a silver-grey iris with a vertical black stripe, creating a distinctive cross pattern on the eye. The colour varies significantly from dark mottled grey flecked with brown or black to chocolate brown or almost white above. Small bright green spots are usually scattered over the back. The colour depends on the amount of light, background colour, moisture, temperature and condition of the frog. Breeding males are more yellowish. The groin and backs of the thighs are yellow with black marbling, and some individuals have dark mottling in the armpit or shoulder area. It is cream or yellowish below, with dark brown flecks on the throat. There are large discs on the fingers and toes. The fingers are half-webbed, the toes almost fully webbed, and the second finger is longer than the first. The skin is rough above with low, rounded warts, and granular below. **Behaviour** This ground-dwelling and climbing frog is active at night, foraging in trees and shrubs, particularly around rivers, creeks and lagoons. It also inhabits areas well away from water, where it lives in trees and forages on the ground on humid nights. It is only active in the warmer months, and hides by day under bark and in hollows and cracks in trees and shrubs. In the breeding season males congregate around waterways and call from branches and other elevated sites, or from the ground if there is no suitable spot, making long, loud, slow cackling calls. **Development** Breeding takes place in spring and summer in semi-permanent ponds and swamps. The eggs are laid in still water and are about 1.5 mm diameter. The tadpoles are quite large with 3 dark longitudinal stripes along the back and a shiny green spot on the tip of the snout. They become olive to yellow brown as they develop. **Diet** Mainly insects and other arthropods. **Habitat** Forests, woodlands, shrublands, grasslands and open sites, usually near rivers, creeks, lagoons and flooded areas, from the coast to drier inland regions. It is often found in trees, around houses and a long distance from water.

LENGTH **To 50 mm**   STATUS **Low risk.**

LEAF GREEN TREE FROG

A small frog with a horizontal pupil and large discs on the ends of the fingers and toes. The second finger is longer than the first, and the fingers have some webbing at the base, while the toes are about three-quarters webbed. The skin is smooth above and granular below, light to dark green above with a narrow golden stripe edged in black running from the nostril through the eye to the flanks. The sides are mottled with white or yellowish blotches. The underside is white, sometimes with darker mottling, and the groin and inside of the thighs are dark red. Individuals may change colour from green to dull purplish-brown, depending on their surroundings. **Behaviour** This climbing and ground-dwelling frog is often seen resting during the day in reeds and among vegetation overhanging streams, swamps and other waterways. It clings to smooth vertical surfaces with the large pads on the tips of its fingers and toes. By night it forages on logs and low bushes near the water. In cool climates it hibernates over winter under rocks and logs close to waterways, emerging in the warmer months. During spring, summer and autumn, males can be heard calling from waterside vegetation by day and night, producing loud 'erk-erk-erk' sounds. **Development** They breed mainly from spring to early summer and females lay loose clusters of eggs in still water attached to sticks and leaves. The small tadpoles are translucent, greyish to light golden brown and metamorphose into young frogs at about 3 months. **Diet** Mainly insects and other arthropods.

Habitat Around creeks, swamps, lagoons and waterholes in rainforests, wet and dry sclerophyll forests, woodlands, heaths and urban areas along the coast and adjacent ranges.

LENGTH **To 40 mm** STATUS **Low risk.**

FAMILY **HYLIDAE**   SPECIES *Litoria raniformis*

GROWLING GRASS FROG
SOUTHERN BELL FROG
GREEN AND GOLD FROG

A large frog with a pointed snout, a horizontal pupil and a golden-yellow iris. There are small discs on the tips of the fingers and toes, and the toes are almost fully webbed. The skin is covered with large warts on the back and is granular on the belly. It is dull olive-brown to bright emerald-green above with gold, black or bronze splashes. A black-edged pale green stripe runs from the nostril, through the eye and down the flanks where it becomes a skin fold. The belly is white while the groin and rear of the thighs are bright blue, sometimes with a few flecks of yellow. Females are larger than males.

Behaviour This ground-dwelling, semi-aquatic frog is active by day and night, and is usually found sheltering among bulrushes and other vegetation at the water's edge. It is a highly mobile species, capable of travelling 1 km in 24 hours. It feeds at night, sitting and waiting for prey to move within range. It often basks in the sun during the day and is seldom seen on roads. It lives mainly in the water, and does not climb into vegetation. Between August and April males can be heard calling while floating among vegetation, making short grunting 'crok-crok' calls followed by a growling 'craw-craw-crawk'.

Development They breed mainly in spring and summer following local flooding and a marked rise in water levels. Females lay up to 4000 eggs in clusters among loose, floating vegetation in still water. The tadpoles hatch 2–4 days later. They are large and swift, yellow to golden, and metamorphose into young frogs when they are about 100 mm long, at 10–12 weeks or the following season.

Diet Beetles, water snails and other small animals it can catch, including small frogs, lizards, snakes and fish.

Habitat Permanent ponds, lakes, lagoons and dams with bulrushes or other emergent vegetation.

Threats Pollution of waterways; habitat loss, fragmentation, disturbance and degradation; predation by introduced fish; fungal disease.

LENGTH **To 105 mm**   STATUS **Endangered.**

FAMILY **HYLIDAE**   SPECIES *Litoria rothii*

ROTH'S TREE FROG

A medium-sized frog of the tropical north, distinguished by the rusty-red colour of the top half of its iris. It has a horizontal pupil and large discs on the tips of the fingers and toes. The second finger is longer than the first and the fingers are half-webbed, while the toes are almost fully webbed. The skin is rough above with low, rounded warts, and granular below. It is light grey to brown above with various amounts of irregular dark mottling, and in direct hot sunlight the skin changes to a uniform pale putty or almost white colour. The underside is cream or yellowish with a single or double dark spot in the groin, visible when the leg is extended. The backs of the thighs are bright yellow or orange with an irregular blackish bar, and there is a black stripe over the base of the forearm. **Behaviour** This climbing and ground-dwelling frog is active at night and avoids dehydration in the dry season and in arid areas by hiding among vegetation, under bark and in tree hollows. It forages on the ground on humid nights, and may be found well away from permanent water, often around houses, relying on puddles, ditches and clay-pans to breed. On the coast it is often found on exposed branches in trees and shrubs overhanging creeks, rivers and lagoons. Males call throughout the summer, particularly after rain, from the ground or from perches in woody vegetation close to water, making a series of 7–9 loud chuckling or cackling calls that resemble maniacal laughter. **Development** They breed in the wet season from November to March in semi-permanent swamps. Females lay around 500 eggs about 1.3 mm diameter in small, free-floating rafts. The pale yellow tadpoles metamorphose into young frogs at 68 mm long when they are around 65–146 days old. **Diet** Mainly insects and other arthropods. **Habitat** It lives in a wide range of habitats, usually associated with larger river systems along the coast, adjacent ranges, and inland areas. It is often found around buildings, in water tanks and drainage systems.

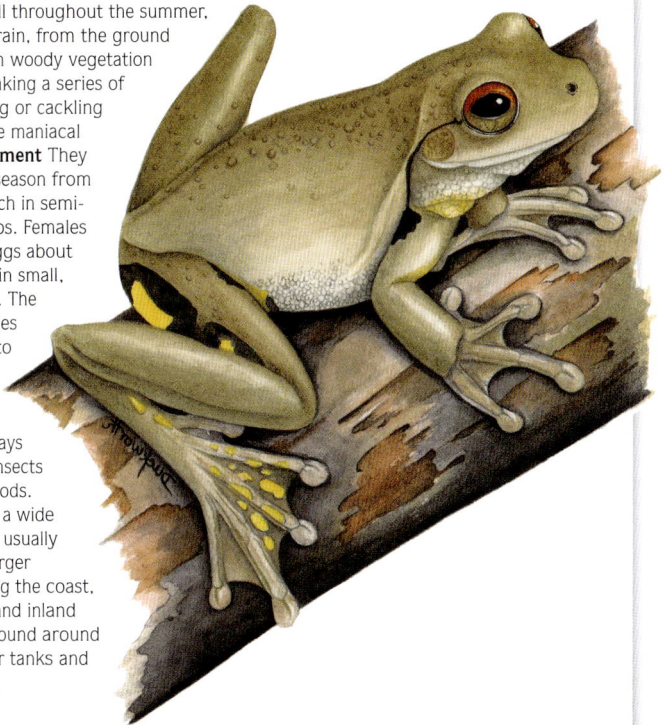

LENGTH **To 57 mm**   STATUS **Low risk.**

FAMILY **HYLIDAE**   SPECIES *Litoria rubella*

DESERT, NAKED OR RED TREE FROG

This small, robust, short-legged frog is one of Australia's most widely distributed species. It has a horizontal pupil and large discs on the tips of its fingers and toes. The second finger is longer than the first and the fingers are slightly webbed while the toes are about two-thirds webbed. The skin is smooth above and coarsely granular below. It is fawn, grey or brown above with a pinkish hue, flecked with black or gold markings. A darker stripe runs down the side of the head, through the eye and along the side of the body, while a broad dark stripe runs along the spine. It is white, cream or yellowish below, usually with a lemon-yellow groin. It changes colour in dry conditions, becoming white to prevent it overheating in direct sunlight. **Behaviour** This ground-dwelling and climbing frog is often found clinging to a window in summer, its internal organs visible through its translucent skin. It forages mostly on the ground at night, and sometimes basks in the sunshine. Its skin is only moderately waterproof, and in arid areas it stays close to permanent water and rests in a water-conserving posture, pressed tightly against a surface with its limbs tucked under its body and its head lowered. Groups of frogs often squeeze into rock crevices to prevent dehydration, or huddle in damp soil under rocks or in tree hollows, changing their position regularly within the group so that each frog spends time in the centre of the cluster where it is warm and moist. Breeding males congregate on the ground or in trees and shrubs beside still or slow-moving water, their loud, screeching, seagull-like calls creating a deafening chorus. **Development** They breed at any time of year in the north, as long as water is present. In other areas they breed after summer rains. Females lay clusters of 40–300 eggs (about 1 mm diameter) in a thin film on the surface of still water. The tadpoles are uniformly brown and metamorphose into young frogs in 2–4 weeks when they are about 35 mm long. The froglets are able to leave the water before they lose their tails. **Diet** Flies, beetles, ants, termites and other arthropods. **Habitat** Most habitats from wet coastal forests to arid areas in the central deserts (confining itself to the ranges or permanent watercourses) and human habitations.

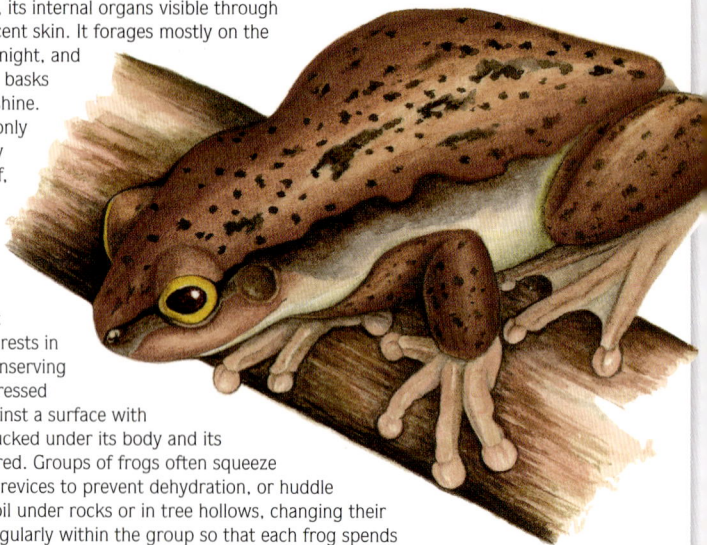

LENGTH **To 35 mm**   STATUS **Low risk.**

FAMILY **HYLIDAE**   SPECIES *Litoria splendida*

MAGNIFICENT TREE FROG

This is one of Australia's largest frogs, and can be distinguished from the green tree frog by the brilliant orange on its limbs, groin and thighs, and the large bulging glands on its head. It has a horizontal pupil and the tips of its fingers and toes have large pads. The second finger is longer than the first and the fingers are webbed at the base, while the toes are three-quarters webbed. The skin is smooth above and granular on the sides and belly. It is pale to dark olive-green above with white, yellowish or orange spots, and white to cream below.

Behaviour A climbing and ground-dwelling frog of the tropical north, it is usually encountered around human habitations where there are cool, moist places. In dry conditions it prevents water loss by secreting a waterproof fluid from its skin, and avoids dehydration by hiding by day in caves and deep rock crevices, gathering in small groups to keep the humidity high. Breeding males congregate around semi-permanent creeks, swamps and pools after heavy rain and call from perches on rocky ledges or from exposed sites, producing low 'crawk-crawk-crawk' sounds.

Development They breed in the wet season from December to January and can remain in the mating embrace for up to 24 hours. Females carry the males on their backs while they lay about 10 separate clumps of some 200 eggs in large floating sheets, distributed over one or more temporary pools, thus increasing their chance of survival. The tadpoles grow to around 55 mm long before they metamorphose into young frogs. These frogs have a life span of at least 25 years.

Diet Mainly insects and other arthropods.

Habitat Rocky sites in the escarpment country of the Kimberley region. It is often found around buildings in water tanks, showers and toilets.

LENGTH **To 120 mm**   STATUS **Low risk.**

| FAMILY **HYLIDAE** | SPECIES *Litoria verreauxii* |

WHISTLING OR VERREAUX'S TREE FROG

A small frog with a horizontal pupil, a distinct eardrum and very small discs on its fingers and toes. The second finger is longer than the first. The fingers are unwebbed while the toes are half-webbed with a small hard pad on the inside edge of the hind foot. The skin is smooth above with small rounded warts, and coarsely granular below. The colour above varies through shades of brown, reddish-brown or fawn with scattered dark blotches. Alpine and sub-alpine populations usually have prominent bright green patches above, while lowland populations have black and yellow markings on the sides, visible when they jump. A dark stripe, sometimes with a pale upper edge, runs from the tip of the snout through the eye and eardrum to the shoulder, and a darker patch or pale stripe often runs down the centre of the back. The front and rear of the thighs are pale yellow to bright orange. The belly is cream-coloured with dark speckles, while the groin area is yellowish with dark brown or black blotches. **Behaviour** This predominantly ground-dwelling frog has a relatively poor climbing ability, and is usually found sheltering during the day beneath logs, rocks and in tussock grasses. It is active mainly at night, although males sometimes call during the day. They call while partly submerged, while floating in the water, or from the banks of dams and swamps in open areas or among low vegetation. Large numbers sometimes congregate during the breeding season. The distinctive high-pitched, repeated, chirping 'weep-weep-weep' calls of the males are heard most of the year in lowland populations, and from spring to early autumn in mountain areas. They hibernate in cold winters to conserve energy. **Development** Mating takes place in still water, usually after rain, between May and November. Females lay 500–1000 eggs (about 1.2 mm diameter) in large jelly-like clumps, and wind them around submerged vegetation. The eggs hatch in late spring or early summer, and the tadpoles grow to about 40 mm long before metamorphosing into young frogs 1–3 months later.
Diet Mainly insects and other arthropods.
Habitat Found around swamps, lagoons and dams along the coast and ranges, in a wide variety of habitats from wet and dry sclerophyll forests to alpine grasslands and bogs.

| LENGTH **To 35 mm** | STATUS **Low risk.** |

FAMILY **RANIDAE**   SPECIES *Rana daemeli*

WOOD FROG

A large frog with long, muscular legs, a narrow snout, a horizontal pupil, and a large distinct eardrum behind each eye. There are very small discs on the tips of the fingers and toes. The fingers are unwebbed, while the toes are fully webbed. The skin is smooth above and below, pale to rich brown above, sometimes with darker flecks or blotches, and whitish below, usually speckled with brown. The lower limbs are light brown with darker cross bars. A dark stripe runs from the snout through the eye to the eardrum and shoulder, and a paler stripe runs below the eye to the forelimb. There are distinct skin folds running along the back from each eye to the base of the leg. Females are larger than males. **Behaviour** This agile, ground-dwelling and semi-aquatic frog is usually encountered at night in or around streams and permanent ponds. By day it often basks in the sun close to waterways, raising its body temperature to aid digestion and increase its rate of growth. It is very wary of predators, particularly birds, and leaps quickly into the water if disturbed, where it stays submerged. The male is unique among Australian frogs in possessing 2 vocal pouches that inflate on each side of the throat when calling. Throughout spring, summer and autumn males call at night from the edge of waterways or from floating vegetation, making distinctive, duck-like quacking sounds. **Development** They breed from February to September, and the female lays large, loose clumps of eggs in swamps and temporary or permanent streams. The tadpoles have gold and black backs and grow to about 60 mm long before metamorphosing into young frogs. **Diet** Mainly insects and other arthropods. **Habitat** Lives in a variety of habitats including rainforests, melaleuca swamps, cane fields, tropical woodlands and seasonally dry monsoon forests

LENGTH **To 81 mm**   STATUS **Low risk.**

FAMILY **GEKKONIDAE**   SPECIES *Christinus marmoratus*

MARBLED GECKO

Australia's most southerly gecko, it is medium-sized and slightly flattened with a long, fleshy tail and long, slender digits with pads at the tips and retractable claws. The skin is smooth and granular, pale grey to pinkish-brown above. Fine blackish lines form a marbled pattern or cross-bands on the back and sides, from the snout to the tail tip. It is whitish below and juveniles often have a row of orange or red spots along the tail. **Behaviour** This gecko is tree-dwelling and rock-inhabiting. It is active at night, particularly in the first few hours after sunset, and shelters by day under loose bark, beneath exfoliating rock and in rocky cracks and crevices. Aggregations of up to 10 individuals are sometimes found, especially in spring. Island populations are less arboreal and usually shelter in cracks and crevices in limestone and granite rocks. The tail is used to store fat, and is readily dropped during aggressive territorial disputes and when attacked. Without its tail the gecko is much lighter and able to run much faster with greater manoeuvrability. **Development** Females lay a clutch of 1–2 brittle-shelled eggs, about 12 mm long, from September to June. They may produce 2 or more clutches in a good season. The eggs are deposited in a nest beneath bark or ground litter and at the base of trees or rocks. One nest was found to contain 40 eggs, indicating its use by a number of females. The eggs hatch from November to April after an incubation period of 2–7 months, and the hatchlings are about 52 mm long with brightly-coloured tails. **Diet** Insects, spiders and other arthropods. **Habitat** Dry sclerophyll forests, woodlands, open shrublands, heaths and rocky outcrops along the coast and semi-arid inland areas, preferring cool moister sites. Also on a number of offshore islands.

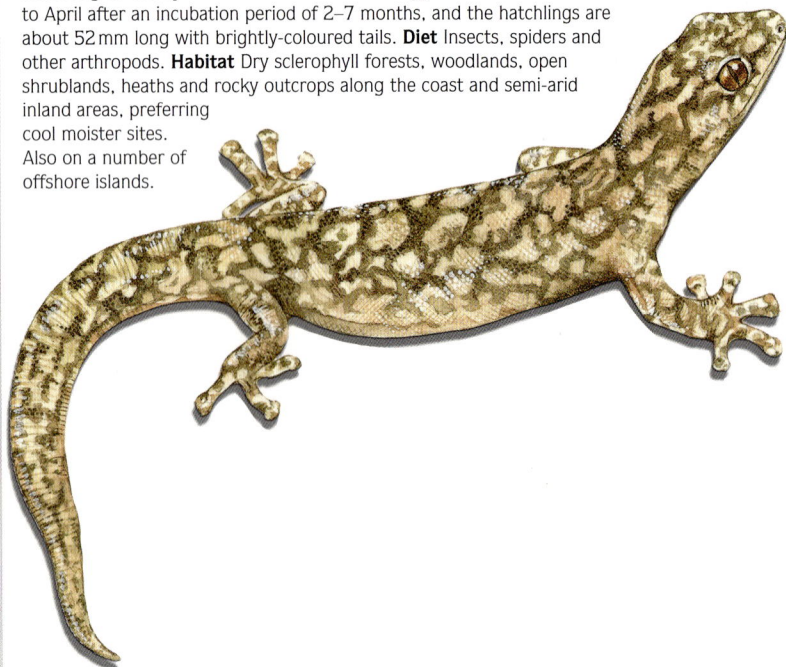

LENGTH **To 145 mm**   STATUS **Low risk.**

RING-TAILED GECKO

This is one of 5 Australian geckos found in northeastern Queensland, previously known as C. *louisiadensis*, which is now confined to Papua New Guinea. It is one of Australia's largest geckos, with a long, slender, prehensile tail and a large, slightly flattened head. It has long limbs with 5 slender, clawed, bird-like digits. It can be identified by 3-4 broad dark brown bands across its back with darker edges, contrasted against a pale brown or whitish background. The bands extend from the back of the head to the tip of the tail. They are darker on the tail, and the first band forms a collar around the back of the head between the eyes. The top of the head is usually mottled with brown and the undersides are white to pinkish. **Behaviour** This gecko is active at night and readily climbs vines, shrubs, boulders, logs and tree trunks looking for food. It also forages on the ground. It is a voracious feeder and a fierce predator, actively stalking, pursuing and ambushing its prey, often launching itself 200 mm or more into the air to catch prey. It shelters by day in caves, and rock crevices, and is sometimes found in buildings. The eyes shining orange-red in a spotlight. It may inflict a painful bite if handled. Mating of this genus has been seen taking place on a branch or log and males have been observed licking the female's face frequently. **Development** They probably mate opportunistically, with a peak from September to March. Females lay 1–2 brittle-shelled eggs, about 11 mm diameter, 3–4 weeks later. The eggs hatch after an incubation period of 3–9 months. The hatchlings are about 128 mm long with brighter colours than the adults. **Diet** Large arthropods including crickets and cockroaches, and small vertebrates including frogs and other geckos. **Habitat** Strongly associated with rocky outcrops along creeks in forested habitats, woodlands and adjacent cleared areas. It is also found in buildings.

LENGTH **To 320 mm**   STATUS **Low risk.**

FAMILY **GEKKONIDAE**   SPECIES *Diplodactylus conspicillatus*

FAT-TAILED GECKO

A medium-sized, fairly stout gecko with a short, thick, slightly flattened, fragile tail that functions as a fat storage organ. Its background colour varies from pale fawn to brown or reddish-brown above, depending on the soil colour, with an irregular pattern of dark lines and blotches, and sometimes pale spots. A pale stripe usually runs from the nostril to the eye. The undersides are whitish with darker patches on the throat. The original tail has rings of large, conical, wart-like projections, although most adults have smoother, more rounded regenerated tails. **Behaviour** This ground-dwelling gecko hunts for insects at night in open areas among grass and rocks, and can occasionally be seen basking in the sun on bare open ground. During the day it usually shelters in ground holes and the empty, silk-lined burrows of trapdoor spiders. It enters its shelter head-first and blocks the entrance with its broad tail. This presumably keeps predators out and helps to keep the gecko moist and cool in the daytime heat. If disturbed at night it runs to its refuge. If it cannot escape, it twists its head to one side and pulls its rear legs up under its body, leaving its tail sticking out. The predator usually grasps the tail which is readily sacrificed by the gecko to avoid being captured. If it is seized, the gecko inflates its body with air to make it harder to swallow, and waves its tail around as a decoy. **Development** Females lay one or more clutches of usually 2 parchment-shelled eggs from October to February. Hatchlings are about 25 mm long. Females are able to reproduce in the season following their hatching and have a lifespan of 3–4 years. **Diet** A wide variety of arthropods, particularly termites and ants. **Habitat** Woodlands, shrublands and spinifex dominated deserts with sandy or hard stony soils, in arid and semi-arid sites along the coast, ranges and inland areas. **Threats** Trampling by livestock and overgrazing by feral and domestic stock.

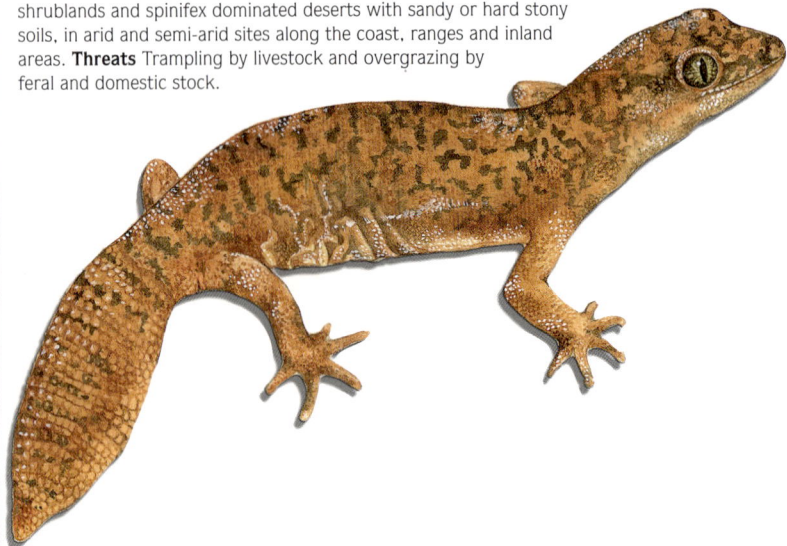

LENGTH **To 110 mm**   STATUS **Endangered in NSW.**

FAMILY **GEKKONIDAE**   SPECIES *Lucasium damaeum*

BEADED GECKO

A small, slender gecko with long limbs, whose digits are covered below by small spines. The skin is smooth and reddish-brown to pink above. A broad, ragged or zigzag-edged, yellowish-brown stripe runs along the spine, and is sometimes broken into a series of blotches. The sides have large cream spots and blotches, usually aligned in a single row. The underside is whitish. **Behaviour** A fast-moving, nocturnal, ground-dwelling gecko, it is very alert and quick to flee if disturbed at night. It shelters by day in the abandoned burrow of another lizard or trapdoor spider, or in the vertical shaft of an ant nest, resting with its head close to the entrance. It sometimes digs a shallow burrow beneath spinifex or other vegetation. It emerges in the first 2–3 hours after dark and forages in open sandy areas, or around sparse shrubs, often covering large distances in its search for food within a home range that overlaps largely with other beaded geckos. Males engage in ritualised combat to establish mating and territorial rights in the breeding season, chirping at each other, raising and arching the body, tail-waving, head-jerking and sometimes biting an opponent's head and body. If disturbed they flee to the nearest cover. **Development** Females lay one or two clutches of 1–2 parchment-shelled eggs between November and January. The eggs hatch in late February. **Diet** Small flying and crawling insects and other invertebrates, including termites, beetles, spiders, crickets and grasshoppers. **Habitat** Savannah woodlands, tall open shrublands, low shrublands, mallee, open scrub, spinifex grasslands and open dunes in semi-arid to arid areas on sand and clay soils.

MOSAIC OR STEINDACHNER'S GECKO

A medium-sized gecko with a relatively large head, slender limbs and a long, slender tail. It has a distinctive pattern of pale, angular blotches and lines on its pale reddish-brown to dark brown back. A pale line runs from above each eye, joins at the back of the neck and continues to the base of the tail, forming 3 large pale patches on the back. The tail has 3–5 blotches, and the flanks sometimes have narrow bars and a line of pale spots. The underside is white. **Behaviour** This ground-dwelling gecko is active by night and shelters during the day in cracks in the soil, suitable holes in the ground(especially in the vertical openings of abandoned insect nests and spider burrows) or beneath surface debris. It emerges after dark and forages at night in open areas between shrubs and other vegetation. This gecko is very alert and flees quickly when disturbed, attempting to confuse its attacker by running in short spurts and then freezing, making it very difficult to catch. **Development** Little is known about their development. Females lay 2 parchment-shelled eggs in a clutch. **Diet** Insects and other arthropods. **Habitat** Dry sclerophyll forests, mallee and savannah woodlands in warm to temperate, semi-arid to arid areas on hilly sites with cracking soil.

LENGTH **To 140 mm**	STATUS **Low risk.**

FAMILY **GEKKONIDAE**   SPECIES *Lucasium stenodactylum*

CROWNED GECKO
SAND-PLAIN GECKO

A medium-sized gecko with long limbs, long, slender digits and almost translucent skin. Its background colour is pink to dark reddish-brown above with pale blotches on the top of the head. Two pale lines run from the end of the snout to the eyebrow, and a pattern of pale spots and sometimes a cream stripe extends from the back of the neck to the tail tip, often broken into blotches. The belly is whitish. **Behaviour** This ground-dwelling gecko is active mainly at night, and is usually seen foraging in open, sandy or grassy areas, generally in low-lying sites and around the lower flanks of sand dunes. It shelters from the daytime heat in the abandoned burrows of other reptiles and spiders, and in soil cracks, where it is cool and humid. In hot weather it licks its upper face and snout rapidly, probably to cool its head. In mild weather it sometimes shelters and forages under surface debris or in spinifex clumps. This is a swift gecko, alert and quick to flee if disturbed. **Development** Females lay 2 parchment-shelled eggs from November to February. Females are able to reproduce in the season following the birth of their hatchlings and have a lifespan of 4 years. **Diet** A wide variety of arthropods, particularly termites and beetles, together with spiders, crickets, bugs and ants. **Habitat** Open savannah woodlands and shrublands with hummock grasses, in sandy deserts and stony areas. **Threats** Habitat disturbance by grazing stock, predation by foxes and cats.

LENGTH **To 90 mm**   STATUS **Vulnerable in NSW.**

| FAMILY **GEKKONIDAE** | SPECIES *Diplodactylus vittatus* |

WOOD OR STONE GECKO

A robust, medium-sized gecko with a short, thick tail, long limbs and long digits. The back is brown to grey with a pale brown, zigzag stripe running along the spine from the top of the head to the hips. The stripe extends along the original tail, while regenerated tails have a number of large blotches or a series of blotches and scattered spots. The underside is greyish or brownish-white, speckled with dark brown. Juveniles are often black above. **Behaviour** This predominantly ground-dwelling gecko is active at night. It rests by day in a burrow dug by itself beneath a rock slab or occupies the burrow of a painted dragon. It also shelters in or beneath fallen timber, below surface debris, under loose bark and beneath small stones. It is sometimes seen at night perching on small branches about 100–200 mm above the ground. In forest habitats it is often found in open areas such as large flat rocky outcrops. It is very tolerant of temperature extremes, remaining active at up to 46°C and penetrating cool, humid areas more than other members of the genus. It hibernates in cold weather, often sharing a shelter with others. If threatened it raises itself on extended legs, inhales deeply, opens its mouth and makes short hisses. **Development** Females lay 2 parchment-shelled eggs about 15 mm long per clutch from December to January. The hatchlings are about 45 mm long. **Diet** A variety of invertebrates including spiders, cockroaches, beetles and other small terrestrial arthropods. **Habitat** Dry sclerophyll forests, woodlands, heathlands and rocky outcrops, from the coast to subalpine regions, in semi-arid to sub-humid areas.

| LENGTH **To 85 mm** | STATUS **Low risk.** |

FAMILY **GEKKONIDAE**   SPECIES *Gehyra dubia*

HOUSE, TREE OR DUBIOUS GECKO

A frequent visitor to houses in northern Australia, this medium-sized gecko has a flattened body, a slender tail and clawed digits with large pads at the tips, giving it great climbing agility. It is pale grey to pale brown above with irregular bands of paler spots, sometimes overlayed with irregular dark lines. The underside is white to pinkish. It becomes paler at night. **Behaviour** This nocturnal gecko lives among trees and rocks. It is swift and agile, and can leap between branches or rocks, and climb up walls and along ceilings. It shelters by day in tree hollows, beneath loose bark, in holes in fence posts and in rock crevices. It has also been found in the disused nests of fairy martins. It emerges at night to forage, usually staying about 2 m above the ground, and frequently enters houses hunting for insects attracted to the lights. Single adult males and females often occur together on the same tree, although individuals of the same sex are never found together. In aggressive displays the gecko arches its back, vibrates its tail rapidly from side-to-side and pecks at its opponent with a closed mouth. If attacked it readily drops its tail and regenerates a new one. It has a soft chattering call. **Development** Females lay 2 brittle-shelled eggs, about 18 mm long, between October and January in the south, and in August or September in the north. The eggs are laid beneath exfoliating bark on a tree stump or its home tree. Several clutches may be laid in a season in northern Qld. The eggs hatch 75–101 days later, and the hatchlings are about 55 mm long. **Diet** A variety of arthropods including flying and crawling insects, spiders, earwigs and cockroaches. They also eat plant material including bananas and sap licked from the trunks of wattle trees. **Habitat** Dry sclerophyll forests, savannah woodlands and rocky outcrops. Often enters human dwellings.

LENGTH **To 145 mm**   STATUS **Low risk.**

| FAMILY **GEKKONIDAE** | SPECIES *Gehyra variegata* |

TREE DTELLA

A small, slender gecko with a slightly flattened body, a relatively long tail and fairly large pads at the ends of its fingers and toes. It is pale brown or grey above with numerous dark blotches and pale flecks. The markings often form a net-like pattern of cross-bands and lines over the back, and there are 2–3 narrow dark lines along the side of the head. The underside is pinkish-white. **Behaviour** This gecko is active mainly at night, generally emerging soon after dusk. It is predominantly arboreal, spending most of its time in trees where it shelters in hollows and under loose bark, preferring shelter sites about 1 m above ground. In rocky areas it shelters in crevices and under exfoliating rock. It also shelters under ground debris and colonises human dwellings. In country towns it is often seen catching insects around street lights at night, and large numbers are frequently found sheltering under sheets of iron and other debris. This gecko forages mainly in the first 3 hours after dark, staying within 10 m of its home site. In cooler weather it basks in the sun and is often active during the day. They live in a home range that may cover 1–6 nearby trees in woodland areas. As many as 11 geckos have been found in the same tree at the same time. In small shelter sites such as tree stumps, males often occur with up to 3 females, but rarely with other males. One or more juveniles often live in stumps with adults for extended periods. If attacked it will readily drop its tail and regenerate a new one. **Development** Females breed in spring and lay a single, brittle-shelled egg, about 11 mm long. Two clutches are probably laid per season, about 1 month apart. The eggs are laid under fallen logs, in hollows in tree stumps, or in holes under rocks. Nests are sometimes communal, and up to 35 eggs have been found in one nest. They are laid between October and early January and hatch after 61–79 days. The hatchlings are about 45 mm long and are able to reproduce in the second season after hatching. **Diet** A variety of arthropods including beetles, spiders, moths, grasshoppers, termites and cockroaches. They also lick sap from wattle trees. **Habitat** Woodlands, shrublands and rocky outcrops from sub-humid to arid areas. It also occurs in human dwellings.

| LENGTH **To 133 mm** | STATUS **Low risk.** |

FAMILY **GEKKONIDAE**   SPECIES *Heteronotia binoei*

BYNOE'S GECKO

This is the most abundant and widespread gecko in Australia. It is small and slender with clawed, bird-like feet and a long, tapering tail. The skin above is finely granular and varies greatly in colour, from pale grey through dull yellowish-brown to bright reddish-brown or almost black. There is usually a prominent pattern of irregular light and dark spots aligned in broad or narrow bands over the back and tail. Most have a dark streak along the side of the head. The underside is whitish to pale pinkish-brown, peppered with darker brown. **Behaviour** A fast-moving, nocturnal, ground-dwelling gecko, it shelters by day in rock crevices, soil cracks, under logs, amongst ground litter, in spinifex clumps, termite mounds and in the burrows of other animals. It moves into cooler, more shaded sites as the sun heats its shelter. Groups of 20–30 geckos are often found under sheets of old roofing iron. It emerges at dusk to forage in open areas, and is very territorial. Each mature male occupies a home range about 10 m wide, which it defends against other males. Threat displays include inflating the throat, arching the back and holding the tail high off the ground while waving it laterally. Fights can last for 30 minutes. An adult male and female often shelter together. Young geckos are tolerated and allowed to live close to adults until they reach maturity and need their own territory. If attacked it readily sheds its tail and regenerates a new one. **Development** Females lay 2 brittle-shelled eggs, about 9 mm long, on the ground beneath bark or logs, in rock crevices or soil cracks, from October to early January. Communal egg-laying sites are used where few good sites exist, with up to 150 eggs, sometimes in layers, being found in one site. They hatch from late February to March. The hatchlings are about 38 mm long and reach sexual maturity at 1–2 years. There are several all-female populations in central and western Australia. They lay unfertilised eggs that hatch into female clones of themselves. **Diet** A wide variety of invertebrates including insects and spiders, and other geckos. **Habitat** Woodlands, shrublands, dry sclerophyll forests and disturbed areas from moist coastal areas to the arid inland, excluding alpine areas and rainforests.

LENGTH **To 120 mm**   STATUS **Low risk.**

| FAMILY **GEKKONIDAE** | SPECIES *Nephrurus asper* |

SPINY KNOB-TAIL GECKO

A large, robust gecko with a distinctive, very short, knobbly tail and a very large, powerful head. It has large eyes with vertical pupils. The limbs are long and slender with clawed digits covered with small spines below. The back and sides are covered with rough, spiny scales. It is pale grey to olive-brown or reddish-brown above, with a number of fine dark bands between transverse rows of pale spots. A net-like pattern of fine dark lines covers the top and sides of the head. It is smooth, pale brown or whitish below, with a rough throat. It can change its colour in a few minutes, generally becoming paler when active and darker when hiding. This is one of the few geckos that does not discard its tail when attacked. **Behaviour** Predominantly ground-dwelling and nocturnal, this gecko is also a good climber. By day it shelters in shallow depressions scraped beneath rocks and logs, in a burrow dug by itself or in the abandoned burrows of other animals, and changes its shelter site regularly. It sometimes lies on a mound to bask in the sun on cold days. At night it forages in open spaces between vegetation, stalking and chasing its prey. It also lies in the entrance to its burrow and ambushes prey that comes close. It is very well camouflaged in colour, shape and texture, and walks in a slow, jerky, swaying fashion, resembling a plant moving in the breeze. It can also swim well, using its legs for propulsion. Individuals sometimes flick dust over themselves, which may help their camouflage. When disturbed or alarmed it inflates its body, rhythmically raises and lowers itself with its forelimbs while twitching its tail from side-to-side, and may leap abruptly and bite its adversary, making a loud, wheezing bark. **Development** Females can store sperm for more than 10 months before fertilising their eggs. They lay up to 5 clutches of 1–2 parchment-shelled eggs, about 27 mm long, from February to April, in a hole dug into moist soil, often beneath a large rock, and backfilled. The eggs hatch about 125 days later, and the hatchlings are about 45 mm long. They have a lifespan of 12 years or more. **Diet** A wide range of arthropods including beetles, spiders, crickets, cockroaches and scorpions, also other geckos and skinks. **Habitat** Open forests, woodlands and sandy heathlands, favouring rocky sites, from sub-humid to arid areas.

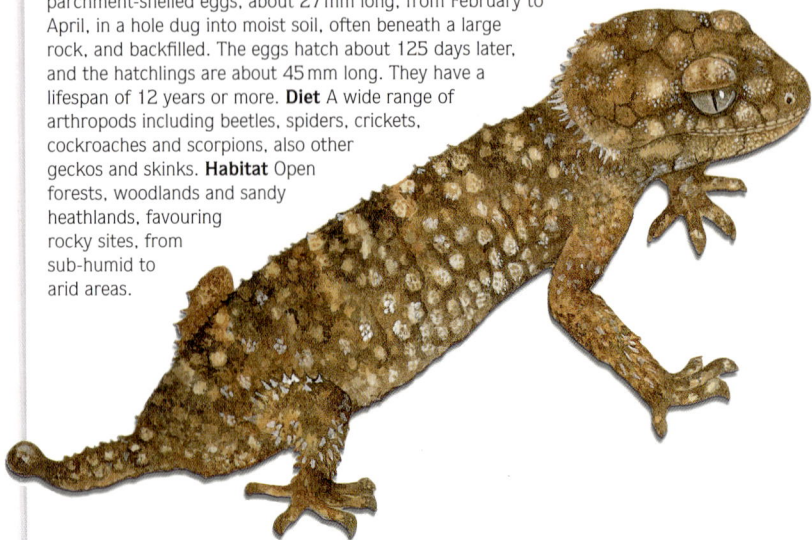

| LENGTH **To 144 mm** | STATUS **Low risk.** |

FAMILY **GEKKONIDAE**   SPECIES *Nephrurus levis*

SMOOTH OR THREE-LINED KNOB-TAIL GECKO

A robust, medium-sized gecko with a short, flattened, knob-like tail and a large head. The limbs are long and slender with clawed digits that are covered with small spines on the undersides. The skin is rough-textured and purplish-brown above. Most individuals have pale, dark-edged bars across the top of the head, neck and shoulders, and small pale spots, bars or lines across the back. It is white below. **Behaviour** This nocturnal, ground-dwelling gecko shelters by day and on cold nights in a burrow dug by itself, or in the burrow of a dragon or other lizard, digging a small side-tunnel into the wall and loosely plugging it with sand. It emerges at night to hunt, particularly after rain, and is much faster at lower temperatures than the smaller geckos it preys upon. It stalks its prey in open areas between shrubs, moving slowly, raised up on its long legs and waving its tail, pouncing when its victim is within range. If threatened it inflates its body with air and pushes itself up and down with its forelimbs in a slow rhythmic fashion while waving its tail. It may leap unexpectedly at its adversary while uttering a loud, wheezing bark. **Development** Females lay a clutch of 2 parchment-shelled eggs, and may lay more than one clutch over much of the year, except in winter. The eggs hatch some 63 days after laying. It has a lifespan of up to 10 years. **Diet** A variety of invertebrates including scorpions, spiders, crickets, cockroaches, caterpillars, centipedes, beetles and occasionally smaller geckos. It obtains most of its water requirements from its prey. **Habitat** Open woodlands, shrublands and spinifex-covered sand plains, tall open heathlands and sand dunes, in arid and semi-arid areas.

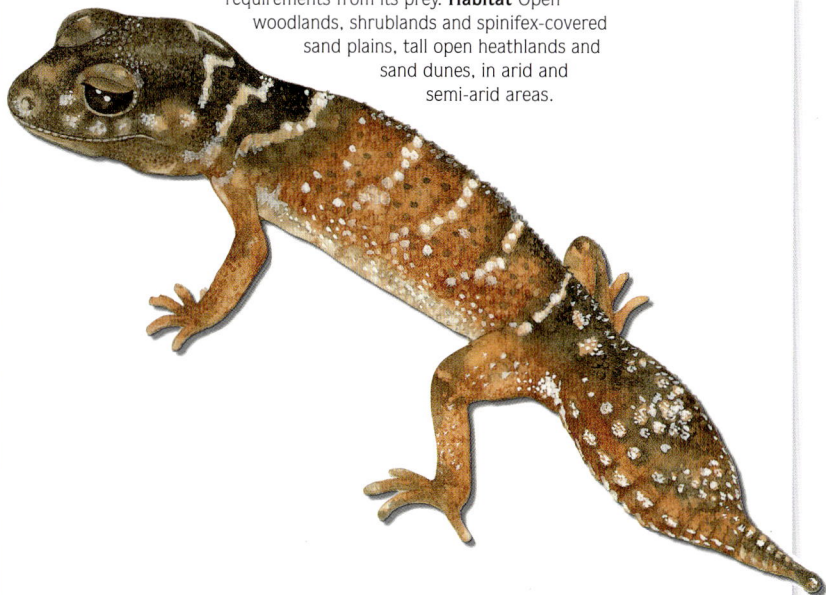

LENGTH **To 115 mm**   STATUS **Low risk.**

FAMILY **GEKKONIDAE**   SPECIES *Oedura castelnaui*

NORTHERN VELVET GECKO

A medium-sized gecko with a flattened body and a plump, flattened tail which functions as a storage organ for deposits of fat. Its long digits have well-developed pads at the tips and small retractable claws. It has a smooth, velvety texture, is deep purplish-brown above, generally with a number of pale bands running from the back of the neck to the base of the tail. The bands are stronger in juveniles and older adults are generally paler overall. The limbs are speckled with yellow and brown, and the belly is whitish. **Behaviour** This gecko is active at night and is an agile climber, spending most of its time in trees. It descends to the ground during the day where it shelters and forages beneath the loose bark of fallen trees, in hollows and among plant debris. It often lives in small communities, and up to 12 individuals may shelter beneath the same sheet of bark. In some districts it lives among rocks, sheltering in cracks and crevices. It forages at night, moving slowly with its body kept low, rising occasionally to leap onto a nearby surface or to escape a predator. If threatened it rises up on its legs and waves its tail slowly from side-to-side, and will readily sacrifice its tail and regenerate a new one. Reserves of fat are stored in the tail when food is plentiful and utilised in poor seasons, allowing the gecko to survive without food for up to 12 months. **Development** They breed from January to September. Females can store sperm and are able to produce 6–9 clutches of 2 parchment-shelled eggs, about 22 mm long, at 3–5 week intervals. The eggs are buried in moist soil and take around 2 months to hatch. Hatchlings have distinctive bars across their backs and are about 60 mm long. They reach sexual maturity at about 18 months. **Diet** Insects, spiders and smaller geckos. **Habitat** Dry sclerophyll forests and woodlands, rocky outcrops and caves.

LENGTH **To 170 mm**   STATUS **Low risk.**

FAMILY **GEKKONIDAE**   SPECIES *Amalosia lesueurii*

LESUEUR'S VELVET GECKO

A medium-sized gecko with a flattened body, a long flattened tail, and long digits tipped with pads and small retractable claws. It has small scales giving it a finely granular to velvety texture, and is pale grey to brown above with a pale, dark-edged, zigzag band along the spine, and numerous small, dark-edged spots or blotches on the sides and limbs. The top of the head and snout are pale with darker mottling. The belly is whitish. **Behaviour** This ground-dwelling gecko is active at night, emerging from its shelter soon after dark and returning several hours before dawn. It is usually found living in caves and rock crevices, where it shelters by day below rock slabs and exfoliating rocks, moving to deeper, cooler crevices in the summer. It forages at night on open rock faces, among ground litter and around the edges of low plants. In some areas it shelters in beehives, apparently coexisting with the bees. **Development** Females lay a clutch of 2 parchment-shelled eggs, about 16 mm long, from December to January. The eggs hatch about 2 months later and the hatchlings are about 46 mm long. **Diet** A variety of invertebrates including cockroaches, crickets and spiders. **Habitat** Dry sclerophyll forests and heathlands near the coast and ranges, on rocky hills and outcrops, particularly sandstone and granite, in well-watered areas.

LENGTH **To 130 mm**   STATUS **Low risk.**

| FAMILY **GEKKONIDAE** | SPECIES *Nebulifera robusta* |

ROBUST VELVET GECKO

A moderately large gecko with a flattened body, a broad, plump tail, clawed digits with pads on the tips, and a velvety texture. It is pale grey to bluish-grey above with a distinct, ladder-like pattern down the middle of the back, comprising pale, squarish blotches with a dark outline. The top of the head is pale bluish-grey, the limbs are finely speckled with dark brown to pale grey, and the belly is whitish.

Behaviour This arboreal gecko lives among trees and around rocks. It is active at night and shelters by day in hollows in tree trunks and limbs, especially those of large, smooth-barked eucalypts, and under the loose bark of dead standing trees. It emerges at night to forage on the lower limbs and trunk of its shelter tree. In south eastern Qld some populations live in sandstone ranges where they shelter in rock crevices and forage on the walls and ceilings of caves. If cornered, it raises its tail and waves it slowly from side-to-side while keeping its head low, before making a sudden dash to its shelter. It also occasionally lives in human dwellings and may lay its eggs behind books and in cupboards.

Development Females lay 2–3 clutches of parchment-shelled eggs in summer. **Diet** Insects and other small arthropods. **Habitat** Dry sclerophyll forests, woodlands, rocky ranges and outcrops, and the fringes of watercourses, along the coast, ranges and slopes.

| LENGTH **To 164 mm** | STATUS **Low risk.** |

FAMILY **GEKKONIDAE**   SPECIES *Oedura tryoni*

SOUTHERN SPOTTED VELVET GECKO

A medium-sized gecko with a slightly flattened, long and slender tail and clawed digits with semicircular pads on the tips. It is chocolate-brown to reddish-brown above, scattered with small cream or yellow dark-edged spots, sometimes forming cross-bars or net-like patterns in older individuals. The belly is whitish. **Behaviour** This gecko lives among rocks and is active at night. It shelters by day beneath flaking rock, in rock crevices, and sometimes under ground debris or under the loose bark of dead standing trees. It lives in small colonies and may share its shelter with up to 7 others, although adult males are never found together. It emerges at night to forage for invertebrates on rock faces in the open. When cornered by a predator it raises its tail and waves it from side-to-side, possibly offering its tail (which will regenerate) as a sacrifice. In cold areas they hibernate for several months over winter. **Development** Mating takes place in September and October. Females lay up to 5 clutches of 2 parchment-shelled eggs, about 23 mm long, at intervals of 29–43 days. The eggs are laid in deep crevices and holes in loose, moist soil, and females often use communal nesting sites where up to 20 eggs may be laid. They hatch about 50 days later in late summer or early autumn. The hatchlings are about 68 mm long. **Diet** A variety of invertebrates, including crickets, moths and other arthropods. **Habitat** Dry sclerophyll forests and woodlands with rocky outcrops, from the coast to the highlands.

LENGTH **To 170 mm**   STATUS **Low risk.**

| FAMILY **GEKKONIDAE** | SPECIES *Phyllurus platurus* |

SOUTHERN LEAF-TAIL GECKO

A medium-sized gecko with a flattened body, a large triangular head and a broad leaf or heart-shaped tail constricted at the base. The limbs are long and spindly with long, slender, bird-like, clawed digits. The skin is granular and spiny, particularly on the tail and flanks. It is well-camouflaged to blend into its sandstone habitat, and is pale grey to rich reddish-brown above, covered with dark brown and black speckles and blotches. Males are smaller than females.

Behaviour This gecko lives around rocks and forages by night around rocky outcrops and nearby forested areas in caves, crevices, on the ceilings of rock overhangs, in rock piles, on tree trunks and in open, stony areas. Its large head helps it subdue large prey, and its flattened body allows it to penetrate into deep, narrow crevices. It frequently enters human dwellings near sandstone outcrops, and may be found in stone cellars and crevices in walls. It shelters by day in deep rock crevices and often shares a shelter site with others. Sixteen individuals were found in one crevice which was littered with their shed skins. If confronted, it makes a loud, long, rasping bark, stands up on its hind legs, gapes its mouth, waves its tail from side-to-side, and may leap forward or turn and flee. If attacked it readily sheds its tail, regenerating a broader, flatter, spineless tail with a shorter tip. **Development** They mate in May and females lay one or more clutches of 2 parchment-shelled eggs, about 23 mm long, from late November to January. They are laid in a deep rock crevice or under a log. Females often use a communal nesting site (one contained 24 eggs) and may use the same site each year. The eggs hatch from January to April, and the hatchlings are about 60 mm long. Females reproduce at 2–3 years of age.

Diet Arthropods, including spiders, cockroaches, crickets, moths, flies and beetles, and juvenile Lesueur's velvet geckos. **Habitat** Wet and dry sclerophyll forests and heathlands, restricted to sandstone areas with caves, crevices and rock exfoliations, along the coast and Blue Mountains. They are often found around gardens in rocky suburbs.

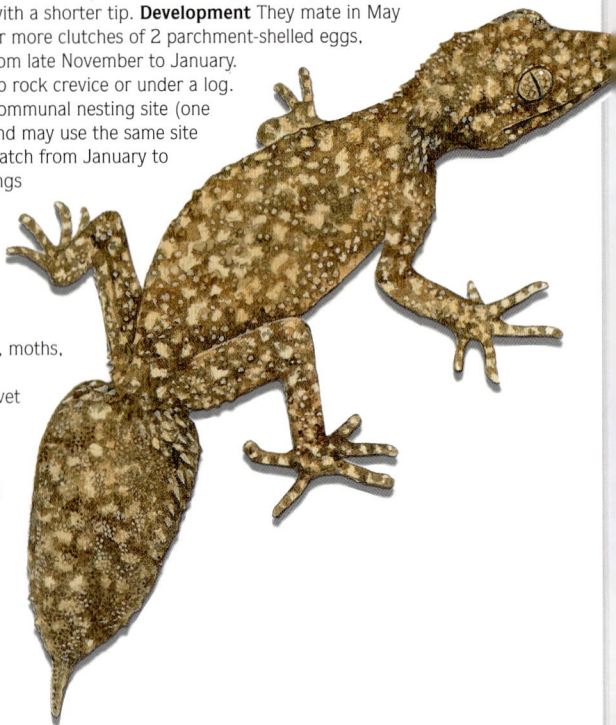

| LENGTH **To 165 mm** | STATUS **Low risk.** |

BEAKED GECKO

A small, slender gecko of the desert regions, with clawed digits, a long, slender tail, a pointed, beak-like snout and large protruding eyes, giving it almost all-round vision. It is red to rich reddish-brown above with dark, irregular lines and small white or creamy-yellow spots over the back, sides and tail, often forming rows. The head is paler grey or brown with a pale stripe along the upper lip and around the eye. It is white below. **Behaviour** This ground-dwelling gecko forages at night in open spaces close to its shelter, but well away from vegetation where the fat-tailed gecko hunts. It is very alert and dashes away quickly if disturbed. During the day it shelters in disused animal burrows (especially those of wolf spiders and larger lizards) and shallow depressions dug under stones or ground litter. On cool nights it often rests with its head protruding from its burrow. In flood conditions it inflates its body to such an extent that it can skim-float across the water propelled by its feet (which do not actually penetrate the water surface), as though it is walking on water. In very hot conditions it licks its face to cool its head. **Development** They mate from October to February and females lay multiple clutches of 2 parchment-shelled eggs. Females are able reproduce in the season following their birth. They have a lifespan of up to 3 years. **Diet** A wide variety of arthropods, mostly termites, some spiders and ants. **Habitat** Savannah woodlands, mallee and mulga shrublands and spinifex grasslands in semi-arid to arid regions with rocky hills, gibber plains, rocky outcrops, sand plains and sand ridges.

LENGTH **To 103 mm**   STATUS **Low risk.**

| FAMILY **GEKKONIDAE** | SPECIES *Saltuarius cornutus* |

NORTHERN LEAF-TAIL GECKO

A large gecko with a flattened body, a narrow neck, a broad, triangular head, and a very wide, leaf-like, spiny tail. The feet are bird-like with long, slender, clawed digits, and the head, body and limbs are covered with conical spines. It is olive-green or brownish above with brown flecks, and usually has 4–5 paler, dark-edged, transverse blotches down the centre of the back. There are a number of W-shaped, dark brown marks on the head, and many individuals have a reddish stripe running along the spine. The belly is whitish to pale olive, sometimes peppered with pale brown. **Behaviour** This gecko lives in trees and around rocks, it is very well camouflaged and is active at night. It shelters by day under peeling bark, in tree holes and in cavities around buttress roots and the roots of strangler figs. In rocky areas it shelters under flaking rocks and in deep crevices. Its flat body allows it to squeeze into narrow spaces where it rests with its head facing down, anchored by the claws of its hind feet. It forages at night on exposed trunks and vertical rock surfaces, and often among the protective leaves of stinging trees. It is active in cool conditions when most other reptiles are resting. Cornered individuals arch their back, raise their tail and slowly wave it from side-to-side, followed by a dash at the aggressor with the mouth agape, making a loud, wheezing bark. If attacked, it readily sheds its tail and regenerates a new, broader, flatter tail. **Development** Females lay a clutch of 2 elongated, parchment-shelled eggs, about 27 mm long, in November and December, in a nest dug into the ground and covered with soil and leaf litter. Females in some areas use a communal nest, and 14 eggs have been found in one nest. Hatchlings, about 70 mm long, emerge 2–3 months later. **Diet** Insects and other arthropods. **Habitat** Warm temperate and tropical rainforests, wet sclerophyll forests and granite outcrops along the coast and adjacent ranges, usually above 750 m.

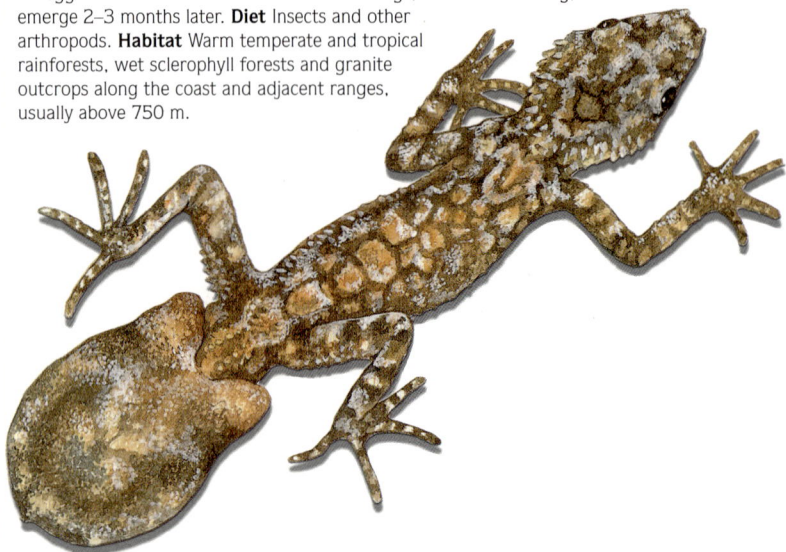

| LENGTH **To 224 mm** | STATUS **Low risk.** |

FAMILY **GEKKONIDAE**   SPECIES *Strophurus ciliaris*

SPINY-TAILED GECKO

A medium-sized gecko with a long, slender body, long spines above the eyes and 2 rows of black or orange spines along the tail. The skin has a rasp-like texture with rows of soft spines or warty projections. Its colour varies from pale silver-grey with scattered black and orange-brown blotches and spots in the NT, through shades of grey, brownish-grey to rich brown mottled with grey, white or orange blotches in other parts of its range. The underside is pale grey to white, usually with dark brown spots. It has large eyes with a grey to brownish-orange iris patterned with fine black lines. The inside of the mouth is dull orange or brownish-yellow. **Behaviour** In forested areas this gecko is predominantly tree-dwelling, while in arid regions some populations live in spinifex clumps. It forages for up to 3 hours after sunset among trees and shrubs, and sometimes on the ground, and is occasionally seen on roads. In warm weather it perches by day and often basks in the afternoon sunshine. In cool weather it shelters in hollow tree limbs, under loose bark, in shrubs or in spinifex grasses. Unlike most other geckos it rarely sheds its tail when attacked. It raises its tail and squirts droplets of sticky fluid from the end of its tail with great accuracy, hitting an aggressor up to 300 mm away. This musky smelling fluid is squeezed from large glands on the upper surface of the tail, and forms irritating sticky threads when it hits its target. **Development** Females breed over a period of up to 9 months and lay 2 sticky, parchment-shelled eggs, about 15 mm long, per clutch. Hatchlings, about 45 mm long, emerge 51–73 days later. They become sexually mature in their third year and may live for 7 years. **Diet** Insects and other invertebrates including crickets, spiders, beetles, caterpillars, cockroaches, grubs, flies, moths and larvae. **Habitat** Open forests, woodlands, shrublands and hummock grasslands from tropical to temperate climates, from humid to semi-arid areas.

LENGTH **To 170 mm**   STATUS **Low risk.**

FAMILY **GEKKONIDAE**   SPECIES *Strophurus elderi*

JEWELLED GECKO

A relatively small gecko with a rather short, plump, prehensile tail and relatively long limbs. It has a background colour of dark brown, grey or black above, and distinctive dark-edged pale spots scattered or aligned over its back. A few spots are also found on the limbs and flanks, and the undersides are greyish-white, scattered with dark flecks.

Behaviour Active at night, this semi-arboreal gecko is always associated with spinifex thickets. It shelters and forages among the spinifex, often in small groups, and only occasionally ventures into open areas. It climbs through the spinifex spines with the help of its prehensile tail, and captive specimens have been observed digging burrows beneath the spinifex to use as a shelter. Like the spiny-tailed geckos, it is able to eject a sticky, foul-tasting, irritating fluid from glands inside its tail. Although capable of squirting the fluid from the end of its tail, it usually smears it over the face or body of an attacker. As a last resort it will drop its tail, which continues to writhe and distract the predator. Such defence mechanisms allow the jewelled gecko to share the same habitat as the gecko-eating Burton's legless lizard, which makes no attempt to attack it, even when no other food is available.

Development Females lay 1–2 clutches of 2 parchment-shelled eggs, about 12 mm long, from September to May. In the southernmost areas they lay their eggs in late January. The eggs are probably buried. The hatchlings emerge some 6 weeks later and are about 35 mm long.

Diet A variety of invertebrates, mainly termites, with some crickets, caterpillars, flies, spiders, cockroaches and moths. **Habitat** Woodlands, acacia and mallee shrublands with spinifex grasses, in semi-arid to arid sand plains and sand ridges. **Threats** Predation by foxes and cats, grazing and trampling by stock, feral goats and pigs, clearing of habitat for agriculture, changes in fire frequency.

LENGTH **To 110 mm**   STATUS **Vulnerable in NSW.**

GOLDEN-TAILED GECKO

This is a medium-sized, slender and strikingly marked gecko. It has a long, slim, partly prehensile tail, long wide digits, a bright orange iris and a vertical pupil edged with white. Its greyish-white back is marked with many irregularly-sized black spots, becoming larger along the spine and denser on the sides of the tail. A distinctive, bright, ragged-edged orange to yellow stripe runs along the tail. The underside is greyish and spotted with black. **Behaviour** The golden-tailed gecko is active mainly at night. It is predominantly arboreal and is often found resting on tree trunks and shrubs with its head pointing down, where it may remain motionless for hours. It shelters in hollow limbs, beneath loose bark, and occasionally beneath fallen logs. This gecko climbs slowly and deliberately among the branches, its body held stiffly above the branch, moving only its legs. If attacked, it smears the aggressor with an irritating, sticky fluid, squeezed rather than squirted from glands in its tail. **Development** Females lay 2 parchment-shelled eggs per clutch. **Diet** Insects and other arthropods. **Habitat** Dry sclerophyll forests and open woodlands, often found among callitris and acacia shrubs, in warm temperate to semi-arid areas.

LENGTH **To 120 mm**   STATUS **Low risk.**

FAMILY **GEKKONIDAE** SPECIES *Strophurus williamsi*

EASTERN SPINY-TAILED GECKO
SOFT-SPINED GECKO

A medium-sized gecko with a long, slender body. It has a partly prehensile tail bearing 4 almost parallel rows of soft, orange-brown spines that extend onto the back. The iris is rimmed with orange-red and has an intricate pattern of fine dark lines. It is pale to dark grey above with a prominent or obscure pattern of dark, wavy lines and black spots. The underside is pale grey with black spots. When hot it becomes a uniform silver grey to reduce heat gain. **Behaviour** This tree-dwelling gecko is active mainly at night, usually in the first hours after dusk. It shelters by day beneath loose bark, in the hollow limbs of dead trees, and sometimes under rocks. In the warmer months it climbs high into the branches and twigs of trees looking for insects, using its tail as an extra limb. In winter it forages closer to the ground, often on tree stumps, and sometimes descends to the ground. In cold weather it hibernates beneath loose bark at the base of trees or stumps. They have broadly overlapping home ranges and return to the same hibernation site each year. It is a sluggish gecko, and often remains quite still for long periods of time. When alarmed it may attempt to frighten its attacker by opening its mouth wide to display the bright purplish mouth-lining and swinging its tail forward. If attacked it squirts the aggressor repeatedly and accurately with an irritating sticky fluid from the tip of its tail. The fluid can be sprayed for a distance of 500 mm and forms cobweb-like filaments. It can also be rubbed on the aggressor. The tail may also be discarded to distract a predator, and a new tail regenerated. **Development** They mate in October and females lay 2 parchment-shelled eggs, about 15 mm long, in November-December, and a second clutch in early January. Clutches have been found in the disused burrows of Gould's goannas. The eggs take 43–48 days to hatch, and the hatchlings are about 43 mm long. They become sexually mature in their first year and have a lifespan of 3–4 years. **Diet** A variety of insects and other arthropods. **Habitat** Dry sclerophyll forests, savannah woodlands and shrublands, especially where callitris and ironbark eucalypts are present, from sub-humid to semi-arid areas.

LENGTH **To 142 mm** STATUS **Low risk.**

FAMILY **GEKKONIDAE**   SPECIES *Underwoodisaurus milii*

BARKING OR THICK-TAILED GECKO

A large gecko with a slightly flattened, plump body and a short, swollen, carrot-shaped tail, constricted at the base, with a slender tip. It has bird-like feet with long, clawed digits. It is deep purplish-brown to reddish-brown above, scattered with small white or yellow spots interspersed with tiny dark spots. These form 2–3 curved lines across the shoulders and the back of the neck, extending to the side of the head. The original tail is black with 5–6 narrow, white cross bands. Regenerated tails are uniformly brown. The belly is white, sometimes with a purplish-brown tinge. Juveniles are very dark purplish-black above.

Behaviour This slow-moving, ground-dwelling gecko shelters during the day under ground litter, below fallen timber, beneath peeling bark, under rock overhangs, in crevices, and in the side chambers of animal burrows (particularly those of wombats and rabbits). It is not very heat tolerant and seeks out deep, cooler shelters in the north. Family groups of 2–6 individuals often share a shelter, and 13 barking geckos have been discovered sheltering under a granite slab in winter. It hunts for and ambushes prey at night in open grassy or sandy sites close to its shelter. It becomes aggressive if disturbed, and when threatened inflates its body and rhythmically pushes itself up and down with its back arched and tail waving slowly. It may leap suddenly at its attacker, uttering a loud, wheezing bark, and will readily sacrifice its tail to a predator. Regenerated tails are rounder, without spines and the slender tip. **Development** They mate in spring and early summer, and females lay one or more clutches of 2 parchment-shelled eggs, about 24 mm long. They are laid in a nest dug in soil beneath a rock or in a deep crevice. Several females may use the same nest. Hatchlings emerge 60–68 days later and are about 50 mm long. Females are sexually mature at 2 years. **Diet** A variety of invertebrates including spiders, scorpions and crawling insects, and also smaller lizards. **Habitat** Wet and dry sclerophyll forests and woodlands, heathlands, mallee, arid and semi-arid shrublands, along the coast and inland, often in rocky country.

LENGTH **To 156 mm**   STATUS **Low risk.**

JACKY DRAGON
JACKY LIZARD

A robust dragon with a long tail and conspicuous eardrum. The back is pale grey to dark brown, becoming lighter when warm. Pale stripes run down the spine from the neck to the base of the tail, sometimes broken into rows of blotches. The tail usually has a series of irregular dark bands. A dark brown bar often runs between the eye and eardrum with narrow dark bands a cross the top of the head. The lips are usually paler than the head, and the mouth lining is bright yellow-orange. The belly is white to pale yellow-brown, sometimes with dark flecks, particularly on the throat. Three rows of sawtooth-like scales run along the back and there are spines on the side of the neck. **Behaviour** This ground and tree-dwelling lizard is active by day, and is abundant in areas regenerating after bushfires and among shrubs and low woodland vegetation. It runs fast on its hind legs, basks on bushes, tree trunks, rocks and logs, and shelters in tree hollows, beneath bark, fallen timber, low vegetation and rocks. It is well camouflaged and ambushes prey, waiting motionless until its victim comes close enough to catch. Individuals use several perching sites in a home range, and deter intruders by displays including tail and arm waving, head-bobbing, push-ups and darkening of colour. Opponents sometimes circle each other with raised, flattened bodies, and may bite each other on the tail. In the breeding season males display from a perch to attract females. They turn dark grey or black, their throat and chest become glossy black, and they drop to the ground to mate. Predators are avoided by freezing, but if cornered or attacked it opens its mouth wide to reveal the bright yellow-orange lining. **Development** They mate in spring and females lay 1–2 clutches of 3–12 soft-shelled eggs, about 11 mm long, between November and January. They are laid beneath logs or rocks, or in a hole 50–100 mm deep, dug in sandy soil and covered with soil and leaf litter. Hatchlings, about 70 mm long, emerge 37–107 days later, from January to March. **Diet** Insects such as grasshoppers and ants, worms, small skinks, berries and flowers.

Habitat Dry sclerophyll forests and woodlands, heaths and sand dunes along the coast and ranges, excluding alpine regions.

FAMILY **AGAMIDAE**   SPECIES *Chelosania brunnea*

CHAMELEON DRAGON

A slender dragon, flattened from the side, with a large head, short limbs, a relatively short tail, and a deep, erectable throat pouch. It has small eyes with scaled eyelids, a conspicuous wedge-shaped eardrum surrounded by a ridge of scales, and a series of furrows on the side of the neck. The head has a low, spiny crest that extends along the back to the base of the tail. It is pale grey to bluish-green above, sometimes with narrow dark brown markings. On the ground it changes colour to reddish-brown or bright yellow. There are dark flecks on the head, 2 narrow brown bars on the neck, and a brown patch behind the eardrum. The tail has a rounded tip and broad, obscure or prominent dark brown bands. The underside is whitish with narrow brown lines from the throat to chest. **Behaviour** This secretive dragon spends most of its time in trees, on debris, logs or low vegetation, and only occasionally descends to the ground to search for food. It is well camouflaged and difficult to spot. If approached it usually does not attempt to flee, but freezes or moves slowly out of sight. If cornered or threatened it displays the broadest side of its body to the aggressor, raises the crest on the back of its head and expands its deep throat pouch. Unlike most other lizards which are fast moving, it has a chameleon-like motion, progressing with slow, deliberate movements, using its prehensile tail to grip branches and objects when climbing, rocking rhythmically from side-to-side. **Development** They mate in trees at the beginning of the dry season and females descend to the ground to lay 2–8 soft-shelled eggs in a shallow burrow from June to September. The eggs are about 21 mm across and hatch about 2 months later. **Diet** Mainly ants and other arthropods. **Habitat** Savannah woodlands.

LENGTH **To 320 mm**   STATUS **Low risk.**

| FAMILY **AGAMIDAE** | SPECIES *Chlamydosaurus kingii* |

FRILLED LIZARD

A powerful dragon with a loose, erectable frill. The frill is yellow to black, often grading to orange or white on the throat, and up to 250 mm across in large males. The colour varies from grey to almost black, brown to brick-red above, mottled with dark brown markings often forming cross-bands on the tail. The belly is off-white to creamy-yellow, and mature males have a pale chest and black belly. NT and WA animals are brick-red with a bright orange and yellow frill, while Qld populations are grey. Juveniles have small frills and are mottled grey and black to blend in with their surroundings. Males are larger than females.

Behaviour This lizard is active by day and is very alert, with acute eyesight. It lives mainly in trees, sheltering in hollows and crevices, spending most of its time in the canopy, particularly in the dry season. It often perches on tree trunks and fence posts, and jumps between branches and to the ground to catch prey. It walks and runs upright on its hind legs with the tail held out as a counterbalance. If disturbed it dashes to the nearest tree or into ground cover. If cornered or threatened it rears up on its hind legs and gapes widely to show its yellow mouth. This action also erects the spectacular frill, making the dragon look intimidating. It will also rock from side-to-side, swing its tail, hiss loudly, and may jump at a potential attacker or lash out with the tail. In the breeding season mature males defend a territory of about 2.5 ha, using their frill as a display, sometimes fighting other males, biting them around the jaw, neck and back. When folded the frill resembles a broken branch or rough bark, and helps camouflage the dragon. The frill may also disperse heat on hot days. **Development** They mate in October and November and females lay 1–2 clutches of 4–23 soft-shelled eggs, about 28 mm long, from December to February. They are laid in a burrow 100–200 mm deep dug into soft soil in a sunny spot and backfilled. Hatchlings, 120–150 mm long, emerge 2–3 months later and stay together for about 10 days. Females begin breeding in their second year. **Diet** Mainly larvae, ants, cicadas, spiders and other invertebrates. Occasionally small lizards, mice, and some plant matter. **Habitat** Dry sclerophyll forests, monsoonal woodlands and semi-arid grassy woodlands.

| LENGTH **To 890 mm** | STATUS **Low risk.** |

RING-TAILED DRAGON

A moderate to large dragon with long limbs and tail, a conspicuous eardrum, and a small crest on the back of the neck continuous with a spiny ridge along the spine. It is pale fawn to orange-brown or bright brick-red above, usually with dark spots on each side of the spine, and irregular cross-bands formed by scattered pale spots and dark flecks. The long, tapering tail is circled with dark brown or black rings. It is whitish or yellowish below, and males have a dark, net-like pattern on the throat, a black patch on the chest and an orange flush on the inside of the limbs. **Behaviour** The ring-tailed dragon is well-adapted to arid areas and is one of the fastest lizards in Australia. It runs on its hind legs with a 'pedalling' action, or on all-fours, chasing prey and leaping to catch low-flying insects. A rock-dwelling lizard, it is active by day and shelters in rock crevices, under large rocks, in hollows or burrows. It basks and displays from elevated perches on rocks or boulders on open, rocky slopes. If disturbed it drops to the ground and runs at high speed into cover, often sprinting a long distance before taking shelter, making it almost impossible for a predator to catch it. They communicate by head-bobbing, doing vigorous press-ups by raising the front of the body while jerking the head up and down, and displaying their throat and chest.

Development They breed in the warmer months from November to March and females lay a clutch of 3–8 soft-shelled eggs. The eggs hatch from January to May. They become sexually mature at about 9 months. Some live to about 20 months, but few live beyond their first year. **Diet** Ants, spiders and other arthropods. **Habitat** Open woodlands, shrublands and hummock grasslands in rocky ranges, slopes, gorges, rocky outcrops and on gibber plains.

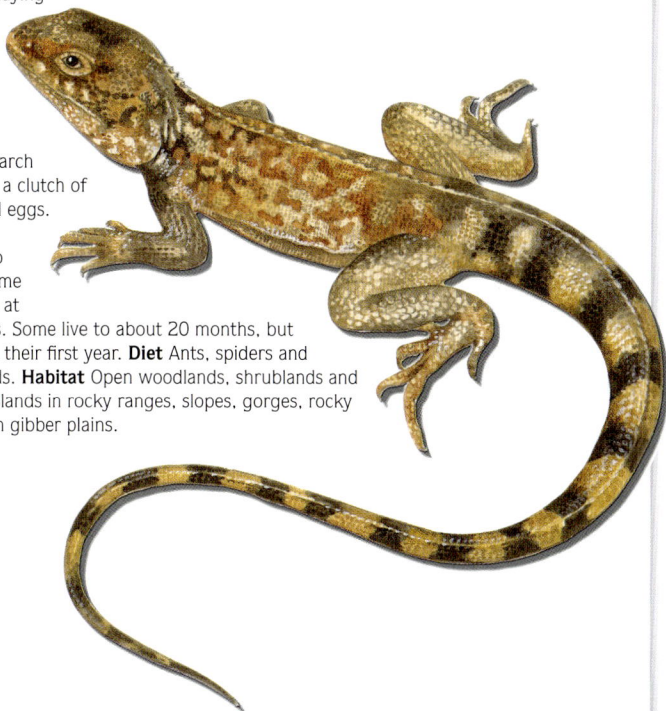

LENGTH **To 350 mm**   STATUS **Low risk.**

FAMILY **AGAMIDAE**   SPECIES *Ctenophorus nuchalis*

CENTRAL NETTED DRAGON

A small, stout dragon with a large blunt head, short limbs and a long, tapering tail. Long spiny scales form a fringe on the lower eyelids and keep sand and dust from the eyes. It has scattered spines on the side of the neck and large, conspicuous eardrums. It is well camouflaged to help conceal it from hawks and other predators, with a pattern of fine dark lines over the body, giving it a netted appearance. The back is pale brown to bright orange-brown or reddish-brown, usually with a pale stripe running from the head to the base of the tail, and a line of black spots along the spine and back of the head. A pale stripe often runs from the eye to the eardrum, and the tips of the claws are pale. It is whitish below, and breeding males are flushed with bright orange on the throat and head. **Behaviour** This ground-dwelling, arid-adapted lizard is active by day, even in very high temperatures. It lives within a home range where it has several burrows, and is often seen along roads, perched on a fencepost, termite mound, rock or log. It runs on all-fours, and is much slower than non-burrowing members of the genus. It digs winter and summer burrows beneath the roots of a shrub or grassy clump, and often plugs the entrance to keep out predators and help control the temperature and humidity. Summer burrows are deep, moist and cool, while winter burrows are much shallower (less than 150 mm deep) so that they warm up enough to enable the lizard to forage above ground around midday. Burrows are often dug into soil thrown up by graders on outback dirt roads and are sometimes shared by 2 adults and juveniles. Displays comprising head-bobbing and raising the upper body from the ground to expose the markings on the chest, throat and neck, are used by males to threaten other dragons and to attract females. If disturbed it flattens itself against its perch or scurries into its burrow. **Development** Females mature early and lay 2–3 clutches of 2–6 soft-shelled eggs in the spring and summer of their first year. Hatchlings (about 65 mm long) emerge after about 11 weeks of incubation. Few lizards survive beyond 2 years of age. **Diet** Mainly ants, termites and beetles, supplemented with other arthropods, flowers and fruits. **Habitat** Semi-arid to arid areas along the coast and inland, usually in open country on red sandy or loamy soils. Often in disturbed habitats, particularly around rubbish dumps.

| LENGTH **To 265 mm** | STATUS **Low risk.** |

FAMILY **AGAMIDAE**   SPECIES *Ctenophorus pictus*

PAINTED DRAGON

Breeding males of this dragon are among the most colourful of all the Australian lizards. They have a blue stripe down the spine, pale orange or yellow flushes on the head and a bright blue flush on the throat and flanks. Out of the breeding season they are bluish-grey to reddish-brown above with a series of black or bluish-grey blotches down the back, interspersed with narrow cream or yellow bands, or rows of pale spots. The flanks have dark patches and scattered pale spots. Females are much duller, usually a dull rusty-brown colour. Both sexes have a small crest on the back of the neck, a ridge of large, erectable scales along the spine and a large, conspicuous eardrum. They are whitish below, and males have a large black patch on the chest and back of the forearm.

Behaviour This fast-moving, burrowing dragon is active during the day, and is often seen perching on branches and fallen logs, or foraging in leaf litter close to its burrow. It digs a U-shaped burrow up to 2 m long beneath low vegetation. The burrow has a concealed entrance and an escape exit, and is sometimes shared by a pair of dragons. The abandoned burrows of monitor lizards, rabbits or wombats are also used. Like many other lizards they signal to each other by head-bobbing, raising the chest off the ground to reveal their chest markings, and jerking the head up and down. Breeding males often confront each other, rising up on their hind legs, with the neck crest and spinal ridge raised and throat distended. Disputes sometimes involve tail lashing and occasionally biting around the head. It relies on camouflage to avoid predators, and freezes if it spots a possible threat. When agitated it raises its crest, and when alarmed it dashes to its burrow or takes cover in ground litter. **Development** They mate in spring and females lay a clutch of 2–6 soft-shelled eggs (about 15 mm long) in a burrow which is then backfilled. The dull brown hatchlings (about 65 mm long) emerge up to 15 weeks later. **Diet** Predominantly ants and other small insects, with some plant matter. **Habitat** Arid and semi-arid woodlands, mallee, chenopod shrublands, hummock grasslands, and around salt lakes.

LENGTH **To 190 mm**   STATUS **Low risk.**

SPECIES *Diporiphora bilineata*

TWO-LINED DRAGON

A slender dragon with a long, thin tail and one or more white spines behind the eardrum. It is grey to reddish-brown above with 2 distinct white or yellow lines running along each side of its body. Between the stripes is a pale grey area often broken with a series of dark brown or blackish blotches that merge into faint bands across the tail. The flanks are pale brown to blackish, usually with pale flecks. It is white or yellow below with pale brown lines on the throat. This species is able to change colour and pattern rapidly to blend in with its surroundings. In the breeding season adult males of some populations develop bright brick-red heads with yellow stripes and black patches on the throat, neck and under their arms. **Behaviour** This largely ground-dwelling dragon is mainly active during the day and is occasionally found on roads after dark. It forages on the ground and perches on low branches in shrubs and trees where it may be found sleeping at night. It also rests in short, shallow burrows. If disturbed or threatened it drops to the ground and runs on all-fours to the cover of grasses or shrubs. If confronted it opens its mouth wide and hisses in an attempt to scare the aggressor. **Development** They breed in the wet season and females may lay up to 2 clutches of 4–8 eggs each year. The eggs hatch in autumn. **Diet** Mainly small insects such as grasshoppers, moths and other invertebrates. **Habitat** Sand dunes, woodlands, wet and dry sclerophyll forests and shrublands with a tussock or hummock grass understorey; along the coast and adjacent hinterland.

LENGTH **To 300 mm**   STATUS **Low risk.**

FAMILY **AGAMIDAE**   SPECIES *Hypsilurus boydii*

BOYD'S FOREST DRAGON

This large dragon lives deep in the tropical rainforests of north eastern Australia and is one of our most spectacular reptiles. It has long hind limbs, a heavy, wedge-shaped head bearing a crest of long white spines that continue along the back to the base of the tail. There is a flat ridge above the eye, a distinct eardrum and enlarged, plate-like scales on the lower corner of the jaw. Below the chin is a drooping yellow throat pouch adorned with a row of large, flat spines. The back is chocolate-brown to purplish-brown or pale grey, becoming black on the sides, often tinged with green. There may be dark bands on the back and sides and cream spots on the flanks. A yellow bar runs between the eye and eardrum. The undersides are pale brown or whitish, and the throat is often darker brown. **Behaviour** Boyd's forest dragon is a slow-moving, inoffensive, tree-dwelling lizard. It relies on its camouflage to avoid predators and is difficult to spot among the foliage. It is usually only seen when foraging along roadsides and around streams. It is active by day and often perches on tree trunks or basks in the sunlight on the forest floor, particularly after rain. It sleeps on exposed trunks or branches in warm weather and uses hollows or dense vegetation for cover in the cooler months. It freezes if disturbed, or moves slowly out of sight, keeping a tree trunk or limb between itself and the intruder. It is territorial, with a home range of about 1 ha and deters other forest dragons from its home range by displaying its colourful throat pouch. **Development** Females lay up to 4 clutches of 3–7 soft-shelled eggs, about 24 mm long, from July to November. The eggs are deposited in a shallow burrow dug beneath a rotten log or in a depression scraped into the forest floor and covered with soil and leaf litter. They hatch around 60–100 days later. **Diet** Rainforest fruits and invertebrates such as slugs and snails, insects, spiders and worms, supplemented with the occasional mouse or bird. **Habitat** Highland and lowland rainforests with vines and smaller trees, along watercourses and forest tracks, to 1100 m.

LENGTH **To 500 mm**   STATUS **Low risk.**

SOUTHERN ANGLE-HEADED DRAGON

A medium to large rainforest dragon with a large angular head
and a pronounced spiny crest on the back of the head. The
spines continue along the back and there are more scattered
spines along the sides and on the upper surfaces of the
limbs. It is grey to grey-brown or chocolate-brown above,
often with a green tinge, although its colour changes
according to the temperature. There may be dark brown, pink
or yellow flecks or spots and dark bands across the back and
tail. A dark band usually runs from the eye to the exposed eardrum, and
there are often dark bars on the jaws. It is whitish or light brown below.
Behaviour This lizard is predominantly tree-dwelling and active by day,
although it sometimes forages on warm nights. It stalks and ambushes
prey in the canopy, on tree trunks, vines, and sometimes on the ground
where it runs on its hind legs, but relatively slowly and clumsily
compared to other dragons. Most sightings are along creeks, roads and
other sunny spots where it forages and basks, particularly after rain. It is
cryptic and inoffensive, but will display aggressively if cornered or
provoked. It avoids predators by blending in with the background and
freezing. They are territorial and males have a home range of
300–700 square metres which they defend by raising their spiny crest,
head bobbing and sometimes fighting other adult males. Males also
engage in ritual combat, particularly in the breeding season, standing on
their hind legs, inflating their body and throat pouch with air, opening
the mouth and circling each other. **Development** Females lay one or
more clutches of 5–8 soft-shelled eggs from November to mid-December.
They are deposited in a shallow depression scraped into loose soil on a
raised bank and partially covered with soil and leaf litter, or in a pile of
leaf litter on the forest floor. Up to 42 eggs have been found at one
site, indicating communal nesting. They are warmed by the
afternoon sun and hatch in late summer after about 72 days of
incubation. **Diet** Mostly stick insects, beetles, insect larvae and
spiders, with occasional worms and small mice.
Habitat Rainforests and adjacent wet forests with vines and
shrubs, along the coast and tablelands, favouring regrowth
areas, watercourses and tracks where there are
canopy breaks.

LENGTH **To 370 mm**   STATUS **Low risk.**

FAMILY **AGAMIDAE** SPECIES *Lophognathus gilberti*

GILBERT'S DRAGON
TATA LIZARD

A large, relatively robust dragon with a slender head, a very long tail and a spiny crest on the back of its neck, extending as a ridge down the spine. It has a large, conspicuous eardrum and scattered spines on the cheek and neck. The back is grey, pale fawn to reddish-brown or almost black. A pale stripe extends from the eye along the sides to the base of the tail, and may change rapidly in intensity, depending on the animal's temperament. They also become darker or lighter to absorb or reflect heat. The back may also be covered with small spots and irregular bars. There is a broad dark band between the eye and eardrum and a white, cream or yellow band from the snout to the rear of the lower jaw. Males develop a black throat and a broad white stripe on the side of the head in the breeding season. Females are duller and slightly smaller with dark flecks on the throat. Southern individuals are usually grey or brown and more muted than northern animals. All are whitish below. **Behaviour** This agile lizard lives partly in trees, and is usually seen perched on a tree limb, termite mound or other vantage point, often near a watercourse. They shelter in tree hollows and among dense vegetation, and sometimes sleep on exposed branches in warm weather. They forage by day, usually sprinting from an elevated perch, but do not actively search for prey. Most movement on the ground is bipedal and if disturbed in the open they dash to the nearest tree, running on their hind legs. After short sprints they bob their heads and wave their arms. Males defend daily activity areas against other males, but not females. Both sexes shift their activity areas on sequential days. Threat displays used by males in territorial disputes include intensifying their colour, raising the crest and spinal ridge, lowering the throat pouch and circling each other with mouths agape. Males butt each other with the head and forebody until one male flees. **Development** They breed in spring and females lay a clutch of 4–8 soft-shelled eggs from November to February. They hatch 59–63 days later. **Diet** Ants, other insects and small lizards. **Habitat** Many habitats from sand dunes to savannah woodlands, mangroves and grasslands, from the coast to the arid interior, usually around creeks and other tree-lined watercourses, also in urban areas.

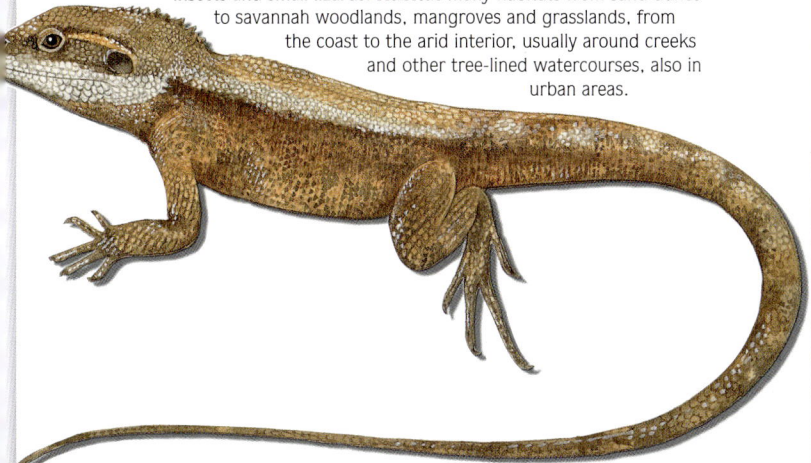

LENGTH **To 425 mm** STATUS **Low risk.**

THORNY DEVIL

This strange-looking, relatively small dragon of Australia's desert regions is covered in large conical spines. It has a globular body with short limbs, a short broad head, a large spiny hump on the back of the neck, a large curved spine above each eye, and a short spiny tail. Generally pale grey-brown to fawn or reddish-brown above, its colour changes to match its background and temperature, becoming paler as it increases in temperature and activity. It is patterned with symmetrical darker markings and reddish patches, and there is a pale stripe along the neck and back. Females are much larger than males. **Behaviour** The sandy colours and irregular shape of this shy and harmless lizard help it blend in with its surroundings. A slow-moving ground-dweller, it positions itself beside ant trails and flicks the ants into its mouth with its tongue. It is active by day, although it avoids the midday heat, and is inactive during the hottest and coldest months. It shelters under low vegetation and in shallow burrows, and forages in open areas, moving 10–500 m daily. It has a jerky motion and often stands with its tail curved over its hips. It survives in arid areas by licking up dew from plants in the early morning and soaking up moisture from damp sand which it kicks onto its back. Tiny channels and folds in the skin act as capillaries, drawing moisture and any dew that condenses on the spines to the corners of its mouth. If attacked it drops its head between its front legs, exposing the fat-filled double-spined lump on the back of its neck to the attacker. During the breeding season males often travel long distances looking for a mate, and may engage in butting contests. Courtship behaviour includes head bobbing. **Development** Females lay 1–2 clutches of 3–10 (usually 6–8) soft-shelled eggs from mid-September to late December. The eggs are deposited in a large chamber at the end of a sloping burrow about 500 mm long and 300 mm deep, dug into soft sand and backfilled. They hatch in late summer, 90–132 days later, and the hatchlings (about 65 mm long) eat the egg cases before digging their way out of the burrow. They mature at 3 years and may live for 20 years. **Diet** Ants, eating up to 2500 per meal. **Habitat** Sand and spinifex deserts and arid scrublands, preferring hummock grasslands, heaths and mallee woodlands.

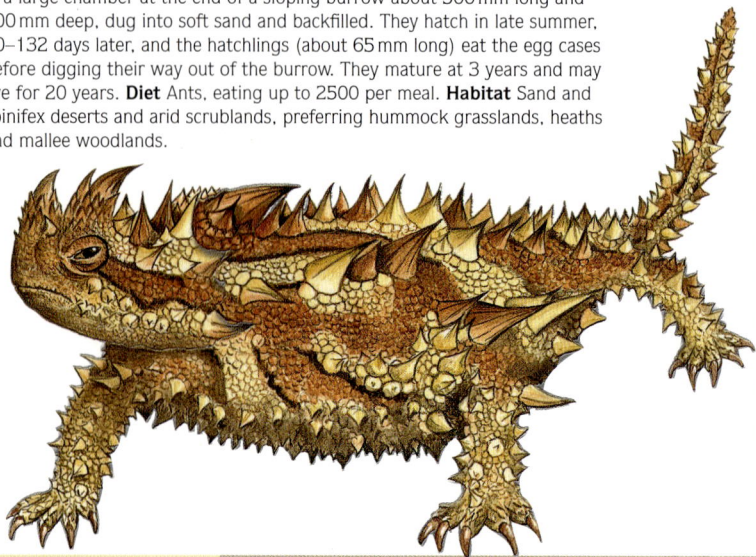

LENGTH **To 200 mm**   STATUS **Low risk.**

FAMILY **AGAMIDAE**   SPECIES *Physignathus lesueurii*

EASTERN WATER DRAGON

Virtually unchanged for 20 million years, the eastern water dragon is thought to be the most ancient of the Australian dragons. It is medium to large with an exposed eardrum, very long limbs and a vertically flattened tail. The back of the head has a crest of raised spines that continue along the back, and irregular spines on the sides, hind limbs and tail. A dark band runs from behind the eye, across the eardrum to the shoulder. There are yellowish bands across the back and tail, and the limbs are dark grey mottled with creamish-yellow markings. The northern form, from central NSW to Cape York, is dark to light brown above, sometimes with darker bands, and has a creamish belly, becoming red to orange on the chest and upper belly in mature males. The southern form is grey-blue above with yellow, orange and blue streaks on the throat, and males develop a blue-green belly. **Behaviour** This semi-aquatic lizard is active by day, and often forages around creeks, rivers and rocky coastal areas. It basks, rests and sleeps on overhanging riverside branches, rocks and logs, and sometimes sleeps in the water. If disturbed it jumps into the water, sometimes from a considerable height. It swims powerfully in a snake-like manner with its limbs held against its sides, and can stay submerged for up to 90 minutes. It is very territorial, and dominant males gather a number of females inside the best territory along a river bank in the breeding season. Other adult males are chased away or attacked. Rivals rear up, attempt to overbalance each other and bite the other's hip region. They can inflict deep wounds with their powerful jaws and sharp teeth, and will do so if handled carelessly. Juveniles congregate separately and old males lead a solitary existence, travelling around looking food, and sometimes scavenging at picnic areas. In cold weather they hibernate in a small, sealed burrow dug beneath a rock or log. **Development** Females lay 1–2 clutches of 6–18 soft-shelled eggs (about 28 mm long) in late spring and early summer. They are laid under cover or at the end of an almost vertical, backfilled burrow 75–110 mm deep. Hatchlings (about 150 mm long) emerge 68–120 days later. They mature in their second year, and may live to 28 years. **Diet** Insects, frogs, worms, mice, smaller lizards, aquatic organisms, flowers and fruits. **Habitat** Woodlands and open forests usually near or along rivers and creeks, next to beaches and in mangrove swamps, from the coastal plains to the slopes and ranges.

LENGTH **To 924 mm**   STATUS **Low risk.**

FAMILY **AGAMIDAE**   SPECIES *Pogona barbata*

EASTERN BEARDED DRAGON

A heavily-built, fearsome-looking dragon with a triangular head, an arc of spines around the neck and cone-shaped scales along the lower eyelids to protect the eyes from sand and dust. Adults have a loose, beard-like pouch on the throat, fringed with long spines. The body is flattened with a relatively short tail bearing a series of spines on either side of the base, and there are strong spines on the flanks to deter predators. It is dark brown to yellowish-brown or pale grey above with a row of pale, semi-circular blotches on each side of the back, sometimes joined to form 2 stripes. Mature males have a black 'beard' and a pale green to blue tinge on the head. The mouth is usually bright yellow inside. The belly is cream with grey circles. Young dragons have more pronounced patterning. **Behaviour** This ground-dwelling and arboreal dragon is active by day, and blends in perfectly with the logs and dead branches it often perches on. It is often seen foraging on roadside verges and basking on fence posts. In the warmer months it sometimes sleeps on a branch, but is more likely to retreat to the shelter of a log, rock or burrow. Winters are spent in torpor in a short sloping burrow loosely plugged with soil. When approached it freezes, or if threatened it attempts to frighten an opponent by standing up, inflating its body and 'beard', and opening its mouth wide to reveal the bright yellow lining. It may rush an intruder, and will bite if handled carelessly. Some 75 different displays have been recorded including head-bobbing, arm-waving, push-ups, head-licking, beard-erection, body inflation, biting and colour changing. Dominant males display frequently, particularly during courtship and in territorial defence. **Development** Females mate in spring and lay up to 3 clutches of 14–31 soft-shelled eggs (about 23 mm long) from August to December in a short burrow dug into moist sand and backfilled. Hatchlings (about 90 mm long) emerge 45–79 days later. **Diet** Mainly insects, but also worms, snails, small lizards, flowers, fruits and other soft plant matter.
Habitat Open woodlands, heaths and dry sclerophyll forests, extending into urban areas.

LENGTH **To 600 mm**   STATUS **Low risk.**

FAMILY **AGAMIDAE**   SPECIES *Rankinia diemensis*

MOUNTAIN DRAGON

This small dragon is the only dragon living in areas with regular snowfalls, and the only one found in Tasmania. It has a blunt head with a large exposed eardrum and a blue mouth lining. Many populations have an orange tongue. It is pale grey to brick-red above with raised spines at the base of the tail and 4 rows of spines along the back. A continuous row of pale semicircles runs along each side of the upper back. Below this a pale line follows a skin fold along the flanks and tail. It is white or cream below. Males are smaller than females and have brighter markings. **Behaviour** The mountain dragon is a fast and active climber, but tends to stay fairly close to the ground among rocks, leaf litter and low vegetation. It is active by day and is often seen in pairs perching on rocks and logs. It shelters among low vegetation and beneath its perching site. It is well camouflaged and rather than pursuing prey it usually ambushes passing insects, remaining motionless until its victim comes close enough to be caught. Large numbers often congregate, particularly in the breeding season, and mature males establish their dominance by rearing up over subordinate males and attempting to drop on top of them, forcing them to retreat or be mounted. Females indicate their willingness to mate by raising their hindquarters and lowering their chest. Other social behaviour includes wrestling, mouth-gaping, head-bobbing and flicking the hind feet. **Development** Females lay 1–2 clutches of 2–7 soft-shelled eggs (about 14 mm long) from October to January. They are deposited in a sheltered site beneath rocks, logs or in a short burrow dug in an open area. Hatchlings emerge in March or April and are about 30 mm long. **Diet** Ants form the major part of the diet, supplemented by a wide variety of invertebrates. **Habitat** Sclerophyll forests, woodlands, heaths and low scrubs on sandy plateaux, stony ranges and rocky outcrops, mainly in highland areas, but reaching the coast in the Sydney region. Forest populations favour clearings.

LENGTH **To 990 mm**   STATUS **Low risk.**

FAMILY **VARANIDAE** SPECIES *Varanus acanthurus*

RIDGE-TAILED MONITOR

A medium-sized monitor with a robust body and relatively short and powerful legs. It has a thick, spiny tail, flattened at the base, with a ridge along the top and one on each side. There are spiny scales over its back and smooth scales covering its head. The back and sides are patterned with numerous yellow spots with dark centres over a dark brown background. There are usually a number of yellow stripes along the neck and side of the head. The tail is dark brown, ringed with yellow or light brown scales. It is white or cream below. Males are much heavier than females with wider bodies. **Behaviour** This monitor lives on the ground and around rocks. It is active during the day, and basks in the sun to warm up. In hot weather it often shelters during the day and hunts in the cooler hours around dawn and dusk. A fast runner and agile climber, it actively hunts for and ambushes small animals, flicking its long, slender and deeply forked tongue in and out frequently to detect the scent of prey. It shelters in rocky crevices, in abandoned termite mounds, in spinifex clumps and in burrows up to 1 m deep, dug beneath rocks, logs or low vegetation. It uses its spiny tail to block the entrance to its shelter, and if a predator tries to pull it out it inflates its body, wedging itself firmly in the crevice or burrow, anchored by its spiny body. **Development** Females lay 1–2 clutches of 2–11 parchment-shelled eggs about 30 mm long in late winter and spring. They are deposited in a nest chamber at the end of a burrow which is back-filled with soft soil. The eggs hatch in early summer, 79–172 days later, depending on the incubation temperature. The hatchlings are about 150 mm long and begin feeding within the first 2 days. They begin breeding at around 18 months. **Diet** Mainly beetles, grasshoppers and cockroaches, with some small lizards, mice and eggs. **Habitat** Woodlands and shrublands with tussock or hummock grasses in rocky mountainous areas and stony plains, from sub-humid to arid areas.

LENGTH **To 630 mm** STATUS **Low risk.**

FAMILY **VARANIDAE**   SPECIES *Varanus giganteus*

PERENTIE

This is Australia's largest lizard and the third largest in the world, weighing up to 20 kg. It has a long neck with an angular brow and a slender tail flattened from the side. It is brown above and densely speckled with cream or yellow spots and blotches edged in black. These are generally aligned in rows around the body and tail, and form a net-like pattern on the side of the head, neck and throat. The pattern is brighter and more contrasting in juveniles and fades with increasing size. The underside is white or cream. **Behaviour** This ground-dwelling monitor is very shy and rarely observed in its habitat. It lives in a deep, secluded burrow among rocky hills, and in cavities beneath boulders. It ventures out in the early morning and hunts in open country in a large home range, often wandering for several days away from its shelter, digging small animals from their burrows with its powerful forelimbs. If it senses any possible threat it lies flat on the ground and remains motionless until the danger has passed. It can run at great speed on all-fours or on its hind limbs, and will readily run up a tree if pursued. If threatened or cornered it raises itself on its hind limbs, distends its neck pouch and exhales noisily with a loud, rattling hiss. If provoked it will lunge and bite and lash out with its tail which is capable of breaking the forelimbs of a dog. In central Australia they are inactive from May to August. **Development** They mate in spring and females lay 6–13 parchment-shelled eggs, about 87 mm long, from November to January. The eggs are deposited in a termite mound or at the end of a burrow up to 1 m long, dug under a large rock and back-filled with soil. Brightly coloured young, about 375 mm long, emerge the following spring. **Diet** Smaller lizards, snakes, mammals (including small kangaroos), birds, eggs, carrion and invertebrates such as grasshoppers and spiders. **Habitat** Woodlands, shrublands, open sand plains dominated by hummock grasses and sand ridges in the arid interior, favouring rocky outcrops in ranges and gorges.

LENGTH **To 2.5 m**   STATUS **Low risk.**

| FAMILY **VARANIDAE** | SPECIES *Varanus gilleni* |

PYGMY MULGA MONITOR

A small monitor with a slightly flattened body. It is light brown to grey above with dark reddish-brown or purple spots often forming bands across the back. The head and limbs have dark brown flecks, and the thick tail has narrow dark brown cross bands at the base and a series of dark narrow stripes on the latter half. The underside is white with numerous grey spots, particularly on the throat.

Behaviour This secretive, tree-dwelling monitor is rarely seen actively foraging. It rests during the heat of the day in tree hollows, under loose bark or in a crack, and emerges at dusk to hunt for small animals and insects living beneath loose bark, occasionally descending to the ground to forage. Prey is swallowed head first. It stays beneath the bark overnight and continues hunting in the cool hours of the morning, returning to a tree hollow when it gets too hot. If cornered it distends its throat, flattens its body and lashes out with its tail. In the breeding season mature males engage in ritualised combat to establish their mating rights. This involves a series of displays including distending the large neck pouch, flattening the body to give a greater impression of size, rising up on the hind limbs, arching the body while supported by the tail and snout, and finally clasping with the forelimbs and hind limbs, vent to vent or side-to-side, and attempting to overbalance the opponent. The victor may bite the loser. Courtship involves head waving by the male, followed by scratching, licking and muzzling the female's neck. **Development** They mate in spring, and females lay a clutch of 2–7 parchment-shelled eggs, about 28 mm long, in late September. The eggs hatch in late December. The hatchlings are about 135 mm long and begin feeding within 18 hours. Females are sexually mature at 2 years. **Diet** Mostly arthropods such as crickets, cockroaches and spiders; some small geckos and skinks. **Habitat** Mulga woodlands and shrublands, often associated with the desert oak, in arid inland areas, on sand plains or sand ridges.

| LENGTH **To 406 mm** | STATUS **Low risk.** |

FAMILY **VARANIDAE**   SPECIES *Varanus gouldii*

GOULD'S OR SAND GOANNA

This is one of Australia's largest lizards. It is light yellow to almost black above with numerous narrow cross-bands and dark-centred spots extending onto the tail. The tail is flattened from the side with a double raised ridge towards the end, and a cream or yellow tip. A black stripe runs behind the eye. East coast individuals are dark with inconspicuous markings, while inland populations are brightly coloured. **Behaviour** Although this monitor lives mainly on the ground it is an agile climber, and sometimes forages in trees and shrubs. It is active by day and shelters at night in a sloping burrow up to 500 mm deep, dug beneath low vegetation, ending in a large chamber with a vertical escape shaft. It often uses several burrows, including abandoned rabbit warrens. It basks in the sun when it emerges, and mainly forages in open areas among grass and surface litter, rising up on its hind legs to gain a better view and sometimes travelling more than 2 km a day. It can run fast and chases prey for long distances, jumping after insects and digging animals from their burrows. If threatened it dashes to its burrow or climbs a tree. If cornered it inflates its neck pouch, rises up on its hind legs, arches its back, flicks its tongue, hisses, thrashes its tail and may lunge at its aggressor. Large numbers congregate in the breeding season and males fight for mating dominance, rearing up on their hind legs, clasping their opponent with the forelegs and writhing on the ground. During courtship the male distends its throat pouch as a display and bites the female on the neck. It becomes inactive from March to August in cooler southern areas. **Development** They mate in spring to early summer and females lay a clutch of 3–20 parchment-shelled eggs from late November to February. The eggs are laid in a long, deep burrow which is plugged and concealed, or under leaf litter, or in a hollow log or termite mound. They hatch the following spring, and the hatchlings are about 260 mm long with bright markings. Females breed at 3 years and may live to 20 years. **Diet** Lizards and their eggs, small mammals, birds, insects, spiders, centipedes and carrion. **Habitat** Most open habitats including heaths, herblands, shrublands and woodlands from the coast to the arid interior.

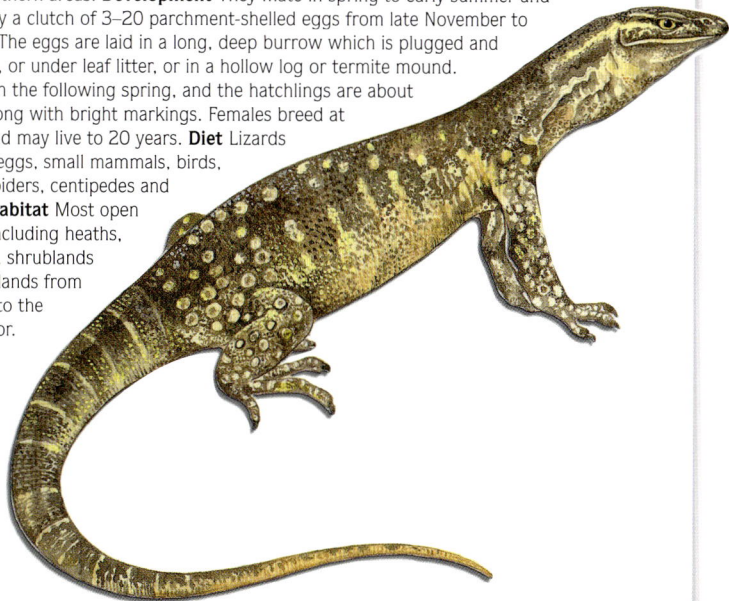

LENGTH **To 1.6 m**   STATUS **Low risk.**

| FAMILY **VARANIDAE** | SPECIES *Varanus indicus* |

MANGROVE MONITOR

A moderately large monitor of the tropical north with a vertically flattened tail. It is dark purplish-brown to black above, covered with small cream or yellow flecks and spots. The underside is white or cream. Females are much smaller than males. **Behaviour** This monitor is well adapted for life on the ground, in trees and in the water. It is active during the day and shelters by night in hollows in tree trunks and limbs. It is a good climber and a powerful swimmer, using its flattened tail to push itself through the water. It forages mostly in forest streams and among tidal mangroves, digging up turtle eggs and preying on fish underwater. It also catches small animals on the ground and in trees. It prefers areas with dense vegetation and is often seen basking on branches above the water. When alarmed it slips quietly into the water or climbs high up in a tree until the danger has passed. In the breeding season males engage in ritual combat to establish mating dominance, rising up on their hind legs, clutching each other with the forelegs, wrestling and sometimes biting and scratching. **Development** Females lay one or more clutches of 1–10 parchment-shelled eggs from August to March. They are deposited in a burrow dug into the soil and concealed beneath rotting wood or ground litter. Hatchlings emerge 152–199 days later. **Diet** Fish, crustaceans, snails, insects, small mammals, birds, marine turtle eggs, juvenile crocodiles, snakes, other reptiles and carrion. **Habitat** Mangroves, monsoon vine forests and tidal waterways, often near fruit bat colonies.

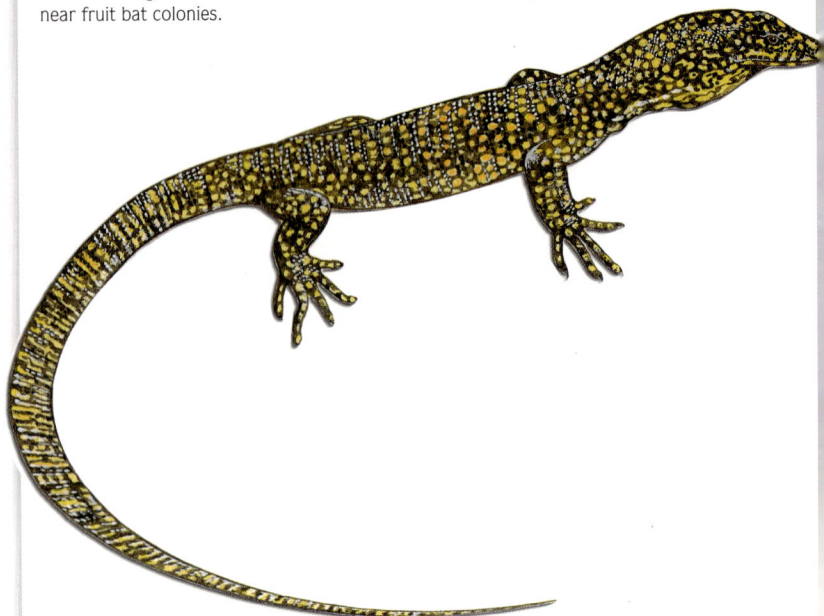

| LENGTH **To 1.5 m** | STATUS **Low risk.** |

FAMILY **VARANIDAE** SPECIES *Varanus mertensi*

MERTEN'S WATER MONITOR

A moderately large monitor of the tropical north. It has a long tail, well flattened from the side with a distinct ridge along the top. The body is olive-grey to dark brown or black above, scattered with small pale spots. The throat is bright yellow and the underside white to yellowish. The nostrils are on top of the snout and flaps of skin automatically close them as they dive. **Behaviour** This semi-aquatic monitor lives among trees and seldom ventures more than a metre from water. It is active by day and frequently basks among aquatic plants. Most of its food is found in the water, although it also forages among the roots of trees lining the bank. In the dry season it feeds on fish caught in receding waterways, sometimes trapping them with its tail. It swims very fast underwater with a crocodile-like motion, propelling itself with its flattened tail, holding its limbs against its sides. It can also walk on the bed of a waterway as though it were on land, keeping its eyes open. It often basks on overhanging branches and on partly submerged timber and rocks. If threatened it drops into the water, sometimes from a great height, and can remain submerged for up to 30 minutes. Alternatively it may climb high up in a tree. Cornered individuals rise up on their hind legs, inflate their throat pouch, arch their back and hiss loudly. In the breeding season males wrestle to establish dominance, standing up on their hind legs, sometimes biting and scratching each other. When food is scarce in the late dry season they burrow into the ground and become inactive. **Development** Females lay a clutch of 3–14 parchment-shelled eggs, about 65 mm long, early in the dry season. They are deposited among leaf mould in a chamber at the end of a near vertical burrow about 500 mm deep, which is firmly sealed after the eggs are laid. Hatchlings, about 300 mm long, emerge in the following wet season, 9–10 months later. They have a lifespan of 20 years or more. **Diet** Mainly fish, frogs and carrion, with some crustaceans, large water insects, lizards, reptile and turtle eggs and small mammals.
Habitat Beside slow and fast moving rivers, often in rocky gorges, around swamps, lagoons and billabongs along the coast and inland.

LENGTH **To 1.3 m** STATUS **Low risk.**

| FAMILY **VARANIDAE** | SPECIES *Varanus mitchelli* |

MITCHELL'S WATER MONITOR

A small, slender monitor with a vertically flattened tail and a pale stripe along the side of the head. It is dark brown to black above, scattered with yellow flecks, spots or black-centred circles. It is cream below and the throat and sides of the neck are bright yellow. **Behaviour** This semi-aquatic arboreal monitor lives among trees and in the water. It is active by day and may be seen around the water's edge, where it shelters and forages among waterside plants, especially pandans. It spends a great deal of time in the water, where it is a proficient swimmer, using its paddle-shaped tail for propulsion. It basks on the branches of trees and pandans overhanging the water, and on partly submerged rocks or tree trunks. At dusk it retires to a hollow in a tree or pandan, or hides beneath loose bark. It is very shy and climbs up the nearest tree or drops into the water if disturbed. **Development** Females lay 3–12 parchment-shelled eggs about 25 mm long in the dry season from April to June. The hatchlings are some 80 mm long. **Diet** Fish, frogs, crustaceans, small reptiles, reptile eggs, nestling birds, mice, insects and other invertebrates. **Habitat** Found in both freshwater and marine habitats along the coast and inland river systems, on the margins of watercourses, swamps and lagoons fringed with trees and shrubs. In the wet season they expand their habitat to include temporarily flooded areas.

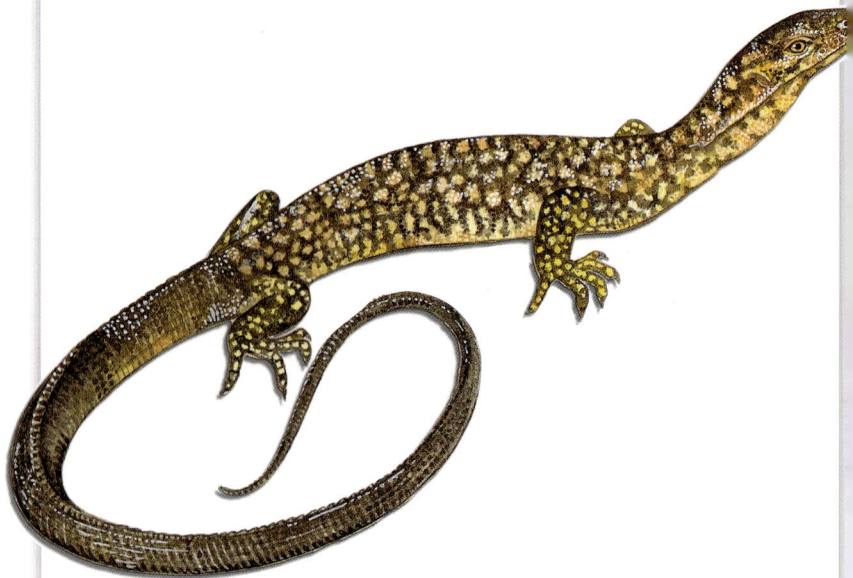

| LENGTH **To 680 mm** | STATUS **Low risk.** |

FAMILY **VARANIDAE**   SPECIES *Varanus scalaris*

SPOTTED TREE MONITOR

A medium-sized monitor with a long cylindrical tail. It is generally grey-brown above with pale flecks and spots. In central Qld they are pale grey, sometimes tinged with rusty-red, and those of the far north east are black above with irregular white or cream cross-bands formed from dark-centred spots. It appears to change colour to match the local tree trunks. The underside is whitish and individuals in the west have a shiny yellow-orange throat. **Behaviour** This tree-dwelling lizard is active by day and is usually seen foraging on tree trunks and branches and occasionally basking in patches of sunlight. It sometimes descends to the ground to feed, and also leaps on prey from as high as 4 m, retreating quickly to its tree. It shelters by night in tree hollows and sometimes under loose bark, emerging to feed at dawn. If disturbed it climbs high into its tree, keeping the trunk between itself and the pursuer. It will also leap to the ground to escape a predator, and climb up another tree. It is active year round, but is seen more frequently in the wet season. **Development** They probably breed early in the dry season, and females lay one or more clutches of 1–10 parchment-shelled eggs about 6 weeks after mating. The eggs are laid in dry soil and loosely covered. Hatchlings, about 165 mm long, emerge 128–155 days later. **Diet** Insects and other invertebrates, mice, frogs, small skinks, geckos, bird eggs and fledglings. **Habitat** Woodlands and tropical rainforests with access to tree hollows.

LENGTH **To 650 mm**   STATUS **Low risk.**

FAMILY **VARANIDAE**   SPECIES *Varanus tristis*

BLACK-HEADED MONITOR

A medium-sized, slender monitor with a cylindrical tail. It is
pale grey to dark brown or black above, scattered with small
white or yellowish spots with dark centres, often aligned
into cross-bands extending onto the tail. The underside is
whitish and populations in WA and central Australia have a
blackish head and neck. **Behaviour** This monitor is found on
the ground, in trees and among rocks. It is active by day, and
hunts early in the morning, warming up quickly by absorbing
heat through its dark head and body. It forages on the ground and in trees,
investigating hollows and crevices and travelling up to 2 km during a day,
returning to the same shelter at night. It is very fast-moving when pursuing
prey or escaping predators, and if threatened it races to shelter in the nearest
rock crevice, burrow or tree. When disturbed it flattens itself on the ground
and arches its tail in the air. Males have a home range of about 40 ha, while
females forage in an area of around 4 ha. They shelter at night in tree hollows,
under loose bark, in rock crevices or burrows, and in some areas take over the
mud nests of swallows and martins. They are active year round in the north,
while in the south they become inactive on cool autumn and winter days. In the
breeding season males follow one or two females about for days until they are
ready to mate. Mating typically takes place in tree hollows. **Development** They
breed from August to October and females lay one or more clutches of 2–17
parchment-shelled eggs, about 33 mm long, from September to November.
They are deposited in a hole about 200 mm deep, dug beneath a tree and
backfilled. The eggs hatch towards the end of summer, 3–4 months later, and
the hatchlings are about 180 mm long. **Diet** Mainly small lizards, birds,
their eggs and nestlings, mice, insects
and other invertebrates. **Habitat** Rocky
ranges and rocky outcrops, in open
forests and grassy woodlands, from
sub-humid to arid areas.

LENGTH **To 800 mm**   STATUS **Low risk.**

FAMILY **VARANIDAE**   SPECIES *Varanus varius*

LACE MONITOR

Australia's second largest lizard, it has a long, slender tail, vertically flattened at the end, with a double ridge on top. It is dark bluish-black above, scattered with white or yellow flecks, spots or blotches. In juveniles these spots are aligned in cross-bands, while in adults they tend to be randomly scattered, except in a few populations in northern NSW and south eastern Qld which have broad yellow and black bands. There are black bands across the snout, chin and throat, and the tail usually has irregular yellow cross-bands. Males are much larger than females. **Behaviour** This lizard is active by day and forages in trees and on the ground, particularly in the afternoon. It basks in the early morning sun and travels up to 3 km a day, investigating hollows and crevices within a home range of around 65 ha. Large animals often scavenge around picnic areas. It shelters by night in tree hollows, in shallow burrows dug below rocks or fallen trees, in termite mounds and rabbit warrens. It uses numerous shelter sites (43 have been recorded), and becomes inactive in cool weather and in winter. If alarmed it dashes up the nearest tree, keeping the trunk between itself and the threat. If cornered it inflates its throat pouch, hisses loudly and may rear up on its hind legs and lash out with its tail. Males often congregate around a receptive female in the spring and summer breeding season. They chase each other and wrestle to establish dominance, rearing up, grasping each other with the forelimbs, and often biting. Courtship involves a dominant male approaching a female and shaking his head vigorously. If the female is receptive she lies prostrate while the male nuzzles and licks her back and sides. Pairs often copulate many times over a few hours, and females may mate with several males.
Development Females lay a clutch of 6–20 parchment-shelled eggs, about 70 mm long, in December or January, in a hole beneath a log or in a tunnel dug into a termite mound. The termites seal over the hole and the hatchlings dig themselves out or are released by the mother about 6–8 months later, in late winter or early spring. They are about 280 mm long and brightly banded with black and yellow.
Diet Nestling birds and eggs, small mammals and reptiles, fish, frogs, insects and carrion. **Habitat** Woodlands, wet sclerophyll forests and rainforests along the coast, ranges and inland.

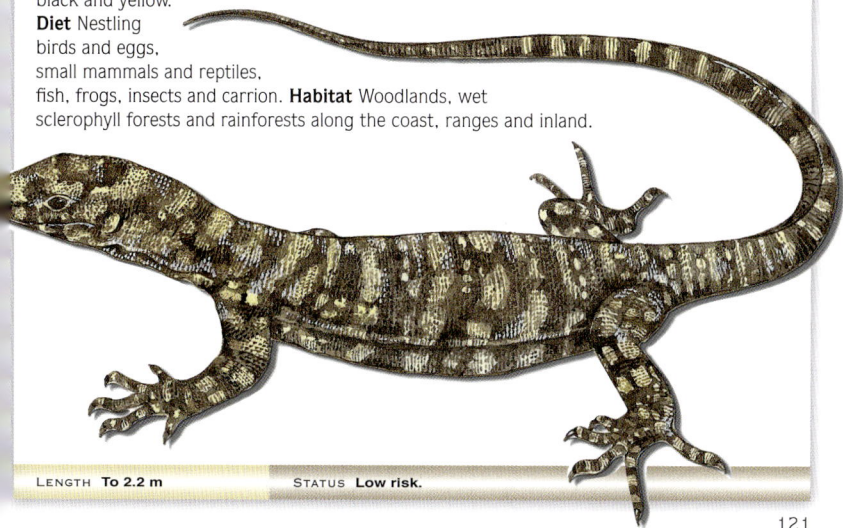

LENGTH **To 2.2 m**   STATUS **Low risk.**

Family **Scincidae**	Species *Acritoscincus platynota*

Red-throated or Fence Skink

A small, relatively robust skink with a fragile tail and short, well-developed limbs, each with 5 digits. It is distinguished by its throat which is bright red in juveniles, orange-red in breeding males and paler in females. Males lose their bright throat colouring outside the breeding season, from late spring to late winter. It is generally pale silvery-grey to brown above with black flecks that sometimes form fine lines. The head is bronze. The upper sides have a broad dark stripe from the snout through the eye to the base of the tail, while the lower sides are silver-grey, merging with the whitish belly. Females are larger than males. **Behaviour** This swift, sun-loving, secretive, ground-dwelling skink is active during the day. It prefers sites with dense ground cover, and is often found in rocky sandstone ridges, sheltering under rock slabs, in rock crevices, in holes and cracks in fallen timber, under logs and beneath leaf litter. It basks on the ground, on fallen timber or rocks near its shelter site, but is very wary and dashes into cover at any hint of disturbance. Females can spot breeding adult males by their bright red throats long before they are close enough to be recognised by their scent. **Development** They mate from late August to mid-October and females lay a clutch of 2–10 eggs, about 12 mm long, in late December or January. The hatchlings emerge in February and March and are about 62 mm long with red throats. **Diet** Mainly ants, other small insects, worms and skink eggs swallowed whole. **Habitat** Lowland and highland dry sclerophyll forests, woodlands, heaths and tussock grasslands with rocky outcrops, favouring west-facing slopes.

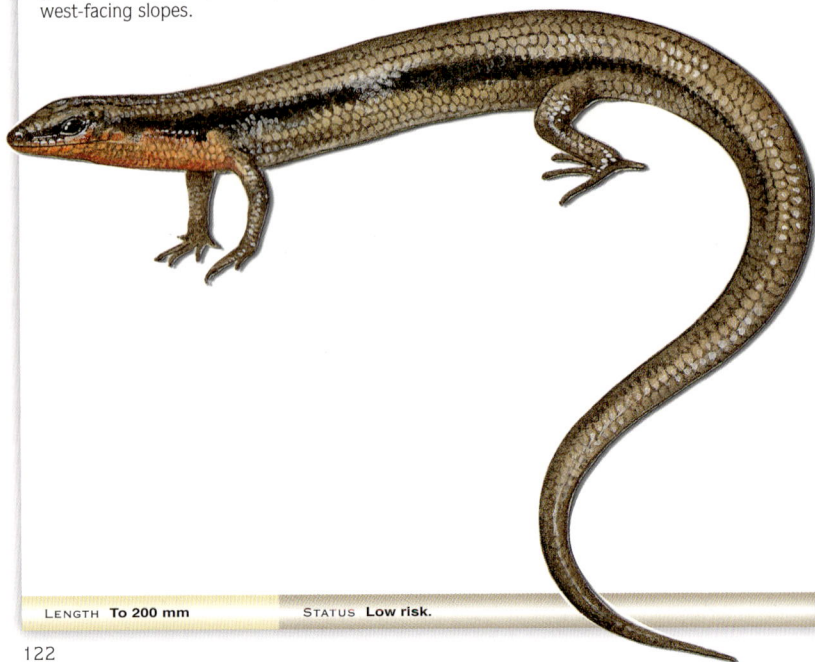

Length **To 200 mm**	Status **Low risk.**

FAMILY **SCINCIDAE** SPECIES *Carlia rhomboidalis*

BLUE-THROATED RAINBOW SKINK

A medium-sized rainforest skink with a conspicuous circular ear opening and well-developed limbs with 4 toes on the forelimbs and 5 on the hind limbs. The top of the head is a rich copper colour, while the back is glossy brown with black flecks, sometimes forming broad dark stripes. A narrow, pale brown to gold stripe usually runs from above the eye to the end of the tail, and there is often a pale stripe running below the eye to the hip. The flanks are blackish with lighter flecks and the belly is white. In breeding males the throat and sides of the neck are flushed with red, or occasionally blue in some southern populations. **Behaviour** This skink spends most of its time on the ground, but it is a good climber and is often seen basking in the sun on low branches in open areas in the forest. It is active during the day, and forages among leaf litter, beneath logs and around tree buttresses in forest clearings. It is fast and waves its tail sinuously from side-to-side as it moves. If disturbed it will dash into cover, and if attacked will readily sacrifice all or part of its tail and regrow a new one. The bright red throat of adult males attracts females in the breeding season. **Development** They mate in September and females lay one or more clutches of 1–2 parchment-shelled eggs under rocks, logs or in dense leaf litter, in a nesting site used by other females. **Diet** Mainly insects with some smaller skinks, including its own species. **Habitat** Rainforests, wet sclerophyll forests and their margins.

LENGTH **To 150 mm** STATUS **Low risk.**

FAMILY **SCINCIDAE**

SPECIES *Cryptoblepharus australis*

INLAND SNAKE-EYED SKINK

A small, flattened skink with long limbs, each with 5 digits. It has snake-like eyes with a clear spectacle over the eye instead of a moveable eyelid. It is grey-brown to dark brown or pale coppery-coloured above with black and pale grey flecks coalescing into bands on the sides and tail. The underside is white or pale metallic-blue and the soles of the feet are brown. **Behaviour** This fast, agile skink is frequently seen on the vertical surfaces of rocks, trees and buildings. It is swift and sure-footed, and active during the day. It shelters at night beneath bark and in cracks and crevices in tree trunks and rocks. It forages in trees and on the ground, actively hunting its prey and snatching food items from columns of foraging ants. It is often seen basking in the sun on fences, in trees, on walls and on rock faces, and commonly inhabits buildings and other human structures. It moves quickly in a stop-start fashion, making it a difficult target for predators. If attacked it readily sheds part or all of its tail and regenerates a new one. **Development** They breed year-round in the north and in spring and summer in the south. Females lay a clutch of 2 parchment-shelled eggs. **Diet** Spiders, beetles, cockroaches, termites, flies, grasshoppers and other insects. **Habitat** Semi-arid zone open woodlands, mallee and acacia scrublands, grasslands and among spinifex. Often found on fences, walls and other human structures.

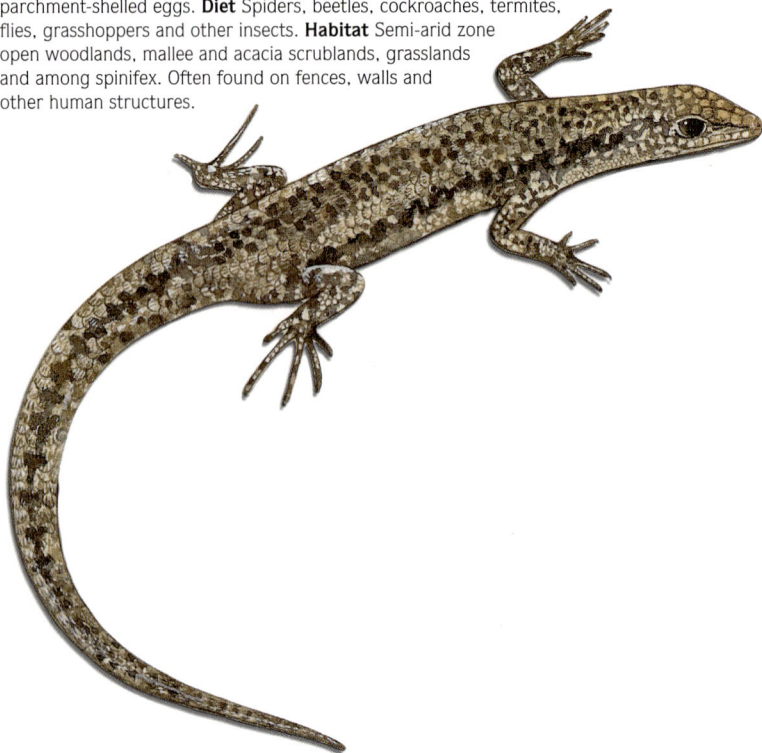

LENGTH **To 130 mm**   STATUS **Low risk.**

FAMILY **SCINCIDAE**   SPECIES *Cryptoblepharus virgatus*

WALL SKINK
STRIPED SNAKE-EYED SKINK

A small skink with a flattened body and long limbs, each with 5 digits. It has snake-like eyes with a clear spectacle in front of the eye instead of a moveable eyelid. It is coppery-brown, silvery-grey to grey-brown or black above, with a narrow cream to yellow stripe on each side running from above the eye to the hips or tail tip. The stripe is bordered by a black stripe above and a broad grey or brown band below. The top of the head is coppery-brown flecked with black. The underside is white or pale metallic-blue. The feet are whitish below and some individuals are completely black above. **Behaviour** This fast moving, semi-arboreal skink is active by day and is commonly found on brick walls and fences in suburban gardens. It shelters at night beneath loose bark, under flat stones and in crevices and holes. Groups of up to 12 have been found sheltering together. It forages among trees and rocks, and occasionally descends to the ground to catch insects or snatch food items from columns of ants. Like other snake-eyed skinks it moves quickly and stops suddenly, making it difficult to catch. If attacked it readily sheds its tail as a sacrifice and regenerates a new one. **Development** They mate in spring and summer and females lay a clutch of 2–3 parchment-shelled eggs from October to late January. The eggs are laid in deep rock crevices, fissures and in low holes in rotting tree trunks. Four females have been observed laying their eggs in a communal nest at the same time. **Diet** Flies, cockroaches, wasps, ants and their larvae, spiders and other small invertebrates. **Habitat** Wet and dry sclerophyll forests, woodlands, mangroves, heathlands, rocky outcrops and urban areas.

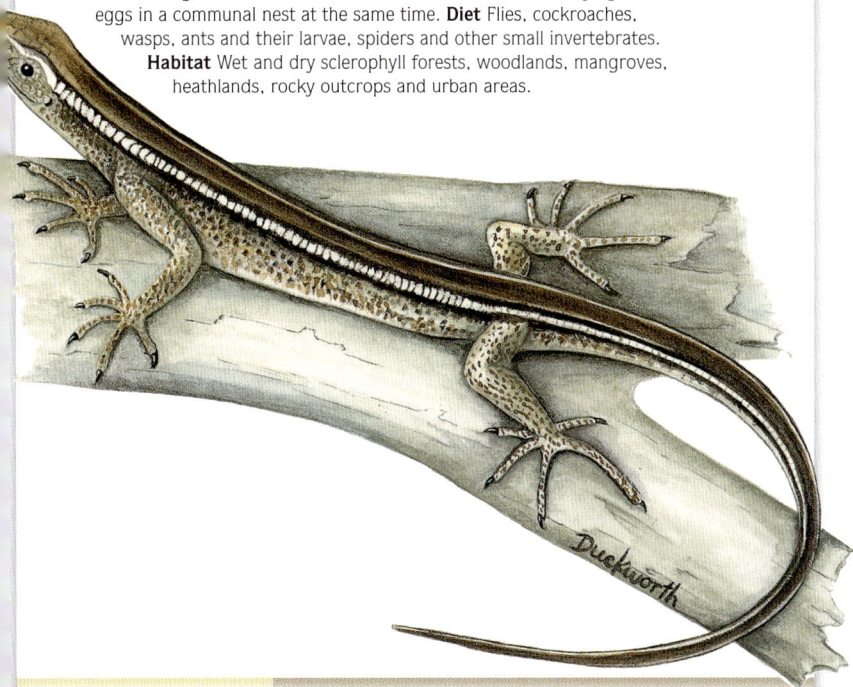

Duckworth

LENGTH **To 100 mm**   STATUS **Low risk.**

LEONHARD'S SKINK

A medium-sized, colourful skink of the arid interior with a complex pattern of stripes and spots. It has long, slender limbs bearing 5 digits, and conspicuous large scales in front of its ear openings. It is pale to dark brown or reddish-brown above, with a dark brown to black stripe along the spine bordered by paler stripes running from the back of the neck to the base of the tail. A whitish stripe bordered by darker stripes extends from above the eye to the middle of the tail, while the flanks are adorned with rows of whitish spots. The underside is white and the limbs are brown with some darker stripes. **Behaviour** This fast-moving, ground-dwelling and sun-loving skink forages mainly in the middle of the day, and it is much more active in summer. It forages widely in open areas searching under fallen logs, beneath tree bark and under ground cover for small invertebrates. It shelters at night in a short burrow dug beneath logs, rocks and low shrubs. If attacked it readily sheds its tail as a sacrifice, and many adults have regenerated tails. **Development** They breed in September and October, and females lay a clutch of 1–7 parchment-shelled eggs from October to December. The hatchlings are 20–30 mm long. **Diet** A wide variety of invertebrates including termites and grasshoppers, larvae, centipedes and spiders.
Habitat Open mulga shrubland, savannah and spinifex areas with low shrubs, in arid areas with stony, sandy or heavy loam soils.

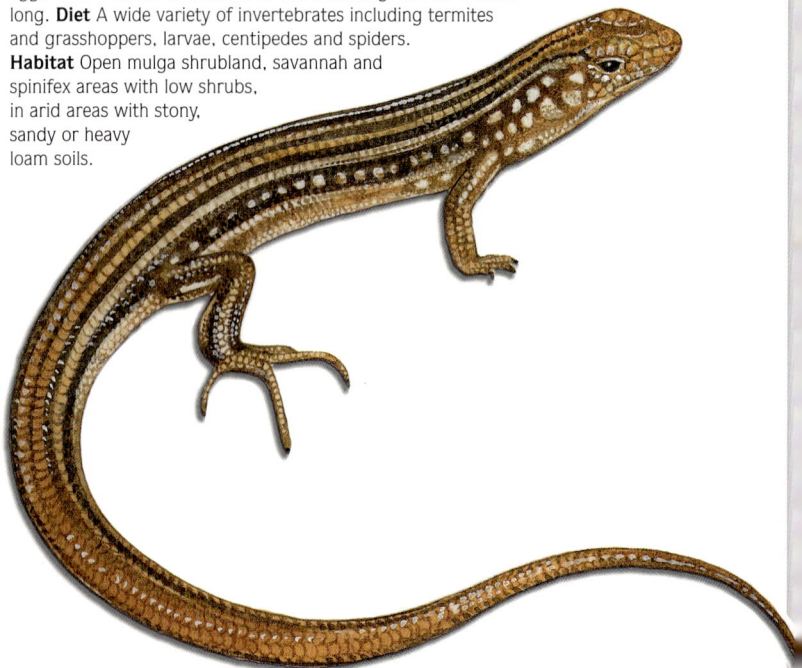

| LENGTH **To 144 mm** | STATUS **Low risk.** |

LEOPARD SKINK

A large, robust skink with a wide body with leopard-like patterning. It has a small head with a wedge-shaped snout and conspicuous large scales in front of its ear openings. It is dark coppery-brown above, usually with a narrow black stripe running along the spine from the back of the neck to the base of the tail. A series of usually 10 rows of black-edged white spots run along the sides and back. The limbs are brown with darker stripes, often with spots on the hind limbs, and have 5 digits. The underside is white. Some populations in the far north have very little patterning. Females are larger than males. **Behaviour** This ground-dwelling skink is active throughout the year during the day and at night. It forages mostly in the cooler hours of the early morning and late afternoon, and after sundown in summer. It shelters in the abandoned burrows of other animals, beneath low vegetation, among spinifex clumps and sometimes under rocks, and forages in open areas between vegetation. It has a distinctive monitor-like gait, moving with a stiff action, holding its body high off the ground. It can move very fast, darting across open ground in straight runs. If attacked it will readily shed its tail as a sacrifice and regenerate a new one. **Development** They breed mainly from August to October and females lay a clutch of 3–11 (usually 7) parchment-shelled eggs from September to January. **Diet** Mostly insects especially termites, with other arthropods including leafhoppers, beetles, spiders and centipedes. **Habitat** Arid and semi-arid areas with sandy loams or stony ground with a low ground cover of spinifex or hummock grasses. Also in open shrublands and woodlands.

LENGTH **To 326 mm**   STATUS **Low risk.**

| FAMILY **Scincidae** | SPECIES *Ctenotus robustus* |

Eastern Striped Skink

A large, widely distributed skink with prominent stripes and conspicuous large scales in front of its ear openings. It is pale to dark brown or olive-brown above with a broad black stripe, edged in white or cream running down the spine from the neck to the base of the tail. A conspicuous pale stripe runs along each side from above the eye to the tail, edged above by a black stripe and below by a broad dark brown band enclosing a number of pale spots or dashes. The limbs are well-developed with 5 digits and streaked, striped or marbled with dark brown. The underside is whitish. Some populations have no patterning and are a uniform fawn-brown above. **Behaviour** This fast-moving, ground-dwelling skink is active by day, even in very high temperatures, and can be seen basking in the early morning sun. It takes refuge in short shallow burrows, usually dug beneath rock slabs, fallen timber or under ground debris. An individual may dig several burrows and move between them during the day. It forages among ground litter, under rocks and logs, and ambushes any prey that comes close to its shelter. It probably has a discrete home range and is very alert, dashing to cover or into the nearest burrow at the slightest hint of disturbance. If attacked, it readily sacrifices its tail and regenerates a new one. **Development** They breed in spring and early summer in southern areas, and females lay a clutch of 2–9 parchment-shelled eggs in a shallow nest, covered with soil. In the tropics females lay a clutch of 3–9 eggs in the dry season. The eggs hatch 46–78 days later and the hatchlings are about 70 mm long. **Diet** Mostly small insects, including grasshoppers, beetles and occasionally small skinks and some plant matter. **Habitat** Lives in a wide variety of habitats from semi-arid to cool wet areas. Open woodlands, shrublands, hummock and tussock grasslands, mallee, savannah, coastal heaths, dry and occasionally wet sclerophyll forests. In the ranges, hills, among rocky outcrops, coastal sand dunes, in granite country and black soil plains.

| LENGTH **To 250 mm** | STATUS **Low risk.** |

FAMILY **SCINCIDAE**   SPECIES *Ctenotus taeniolatus*

COPPER-TAILED SKINK

A medium-sized skink with a distinctive pattern of contrasting stripes and a long and pointed, brown or coppery-orange tail. Like other *Ctenotus* skinks it has conspicuous large scales in front of its ear openings and well-developed limbs with 5 digits. It is rich brown above with a series of dark and pale stripes running along the back and sides, from the snout to the base of the tail. A black line also runs along the side of the tail to the tip. The limbs are reddish with black streaks, and it is whitish below. Animals living in coastal sandy sites often have a pale olive tail. Females are larger than males. **Behaviour** This fast-moving, ground-dwelling skink is active by day and is often seen basking on a large flat rock. It shelters in a burrow dug beneath a rock, and in coastal sandy areas it digs shallow burrows below low shrubs. It also shelters in crevices, under rock slabs or under logs, and several individuals are sometimes found sheltering under small rocks in rocky outcrops. It forages in open areas and around low shrubs, poking its snout into holes and digging with its front feet. It dashes into cover at the least hint of disturbance, and if attacked will readily sacrifice its tail and grow a new one. **Development** They mate in spring and early summer, and females lay a clutch of 1–7 parchment-shelled eggs in a nesting burrow dug into sandy soil beneath a rock. The hatchlings emerge in January after an incubation period of around 40–66 days, and are about 60 mm long. They become sexually mature at 1–2 years. **Diet** Worms, beetles, spiders, millipedes, wasps, ants, flies and other invertebrates. **Habitat** Dry sclerophyll forests, woodlands, heathlands and shrublands on stony or sandy soils, usually with rocky outcrops or rock slabs. Also on sand dunes behind beaches and disturbed sites that are regenerating.

LENGTH **To 200 mm**   STATUS **Low risk.**

FAMILY **SCINCIDAE**   SPECIES *Egernia cunninghami*

CUNNINGHAM'S SKINK

A large, robust skink with a plump body and a thick, spiny, fragile tail. The scales on the back and sides each bear a short, sharp spine that becomes longer and much more pronounced on the tail. It is commonly grey-brown to olive-brown above, scattered with lighter flecks. The eyelids are white-edged and the belly is usually white. There are several isolated populations varying in size and appearance. In the far north they are very dark brown to black with little patterning, and pink to orange below. Those from northern NSW and south eastern Qld have a prominent pattern of cream to white blotches forming bands across the back and tail, and are whitish below with dark brown markings on the throat. More southerly individuals are generally paler with a salmon pink or cream belly. The limbs are well-developed with 5 digits and the fourth toe is much longer than the third. **Behaviour** This ground-dwelling and rock-inhabiting lizard is active by day and around dusk in hot weather. It is gregarious and lives in small colonies or family groups throughout the year (a mixed colony of 17 males and females has been found in one site). It shelters in rock crevices and beneath rock slabs, and solitary individuals are occasionally found living in hollow logs or cracks in dead trees. Members of a colony mark their inhabited rocky outcrop by using a nearby rock as a common defecation site. It is very alert and retreats rapidly into cover if disturbed. If any attempt is made to remove it from its shelter the lizard inflates its body with air so that its spiny scales grip onto the sides of the cavity, making it almost impossible to extract. **Development** They mate in spring and females bear a litter of 1–11 live young in mid to late summer. The newborn are about 120 mm long. They are sexually mature at 5 years and may live to 30 years. **Diet** Adults eat mostly native berries, soft leaves and shoots, with some insects, spiders, worms, snails and small lizards. Juveniles are carnivorous. **Habitat** Open forests, woodlands, shrublands and heaths with cool rocky outcrops, in ranges, hills, river valley slopes and coastal cliffs.

LENGTH **To 450 mm**   STATUS **Low risk.**

PYGMY SPINY-TAILED SKINK

A relatively small, robust skink with a very short, spiny tail and a flattened body. It has strong, well-developed limbs and the fourth toe is longer than the third. The scales on the tail each bear 3 spines and the scales on the back each have a prominent spine and 2–4 sharp ridges. It is pale brown to bright reddish-brown above with obscure darker cross-bands. Northern individuals have irregular black spots or blotches forming bars across the body and tail. Those in the south tend to become brownish-grey from the mid-body, with irregular brown blotches. The flanks and limbs are pale brownish-grey with black blotches, and the underside is greyish-white, often with darker flecks on the throat and chest. **Behaviour** This lizard lives mostly on the ground, but spends some of its time in trees. It is active during the day and lives in small colonies or family groups in tree hollows, rock crevices and large termite mounds. Colonies mark their home site by defecating in the same spot. In the breeding season females chase other skinks from their home shelter where they bear their young. They avoid predators by running to a crevice or hollow and blocking off the entrance with their spiny tail. If an attempt is made to pull it out, the skink inflates its body with air, jamming itself into the hole with its sharp spines gripping the sides, making it almost impossible to dislodge. **Development** They breed in spring and females give birth to 2 live young in summer. **Diet** Primarily insects, spiders, grubs, centipedes, with some flowers, fruits, soft leaves and shoots. **Habitat** Semi-arid to arid sites with rocky ranges, hilly areas and granite outcrops. Often found in mulga woodlands and other arid shrublands.

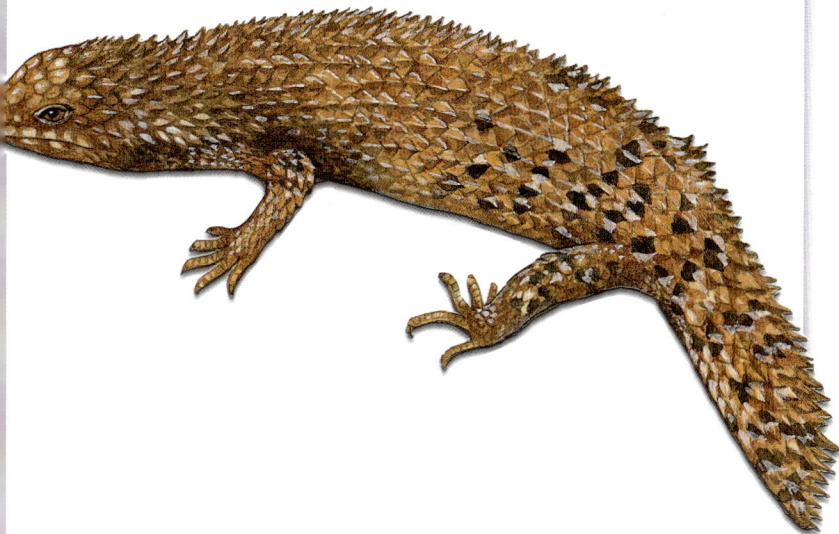

LENGTH **To 161 mm** STATUS **Low risk.**

FAMILY **SCINCIDAE**   SPECIES *Egernia frerei*

MAJOR SKINK

A large skink with a flattened body and smooth scales. The limbs are well-developed, with 5 digits, and the fourth toe is much longer than the third (although most adults have several missing toes). It is pale yellowish-brown to rich brown above with fine dark lines from the neck to the base of the tail. These markings are much more prominent in individuals living on the east coast, and usually form 3–5 dark lines along the back. Northern populations have a much weaker pattern. The sides are dark brown to blackish with scattered whitish spots. The underside is white, pale grey to orange, and is often paler on the throat and tail. **Behaviour** This shy, ground-dwelling skink is active by day, and is usually seen basking in the sun at the edges of clearings and along tracks. It forages in or around thick ground cover, vines, fallen timber, tree buttresses, and among vegetation around rocky outcrops. It is fast-moving, very wary, difficult to approach, and dashes to cover if disturbed. It shelters in hollow logs, in cavities in the root systems of fallen trees, or in burrows beneath large rocks or logs. It either digs its own burrow or takes over the abandoned burrows of other animals, and may live in small colonies. **Development** They mate in spring and females bear up to 3 live young, about 11 mm long, in summer. They have a lifespan of 11 years or more. **Diet** Insects, spiders, lizards, worms, snails, mice, fruits and soft plant matter. **Habitat** Rainforests, vine thickets, wet and dry sclerophyll forests and woodlands, usually around rocky outcrops in coastal and near coastal areas.

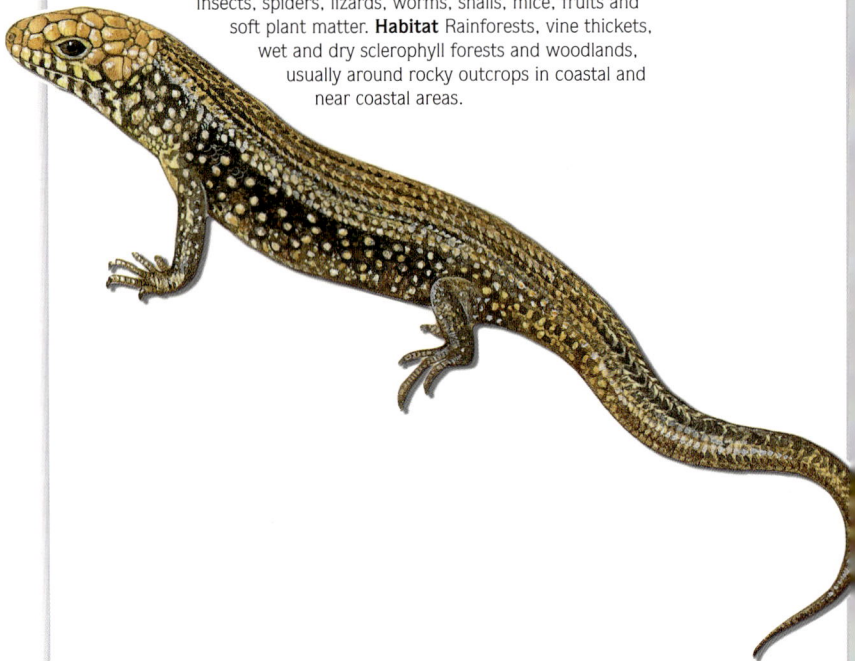

LENGTH **To 470 mm**   STATUS **Low risk.**

FAMILY **SCINCIDAE**   SPECIES *Liopholis inornata*

DESERT SKINK

A relatively small, stout, smooth-scaled skink of the arid regions, with a large blunt head and a short tail. It has short legs with 5 digits, and the fourth is much longer than the third. It is yellowish-brown to rich coppery-red above, depending on the background colour, often with fine dark lines and scattered black spots forming vertical bars on the flanks and tail. The underside is white. **Behaviour** This ground-dwelling lizard is active mainly in the late afternoon and evening, and appears to be solitary. It shelters in a complex burrow system dug into sandy soil at the base of a shrub or spinifex clump. The burrow is around 500 mm deep and 1.8 m long, with a deep resting chamber and up to 5 narrow tunnels. Each tunnel has its own entrance covered with a thin crust of sand or is hidden among shrubs or spinifex. These are used as ventilation shafts and escape routes if necessary. They warm their bodies near the surface of these tunnels, avoiding exposure to predators. The entrances usually face north to north west, and individuals may use 2 burrows 10–20 m apart. The desert skink often rests with its head protruding from its burrow, probably looking for passing prey. In late autumn, activity centres on entrances receiving sun, and in winter it retires to a sealed-off section of the burrow and becomes inactive. **Development** Females give birth to 1–4 (usually 2) live young between September and early May. **Diet** Mainly arthropods including ants, spiders and centipedes, also smaller lizards and plant matter. **Habitat** Mallee woodlands and shrublands with hummock grasslands on arid and semi-arid sand plains and sand ridges.

LENGTH **To 200 mm**   STATUS **Low risk.**

LAND MULLET

This is Australia's largest skink, and one of the largest skinks in the world. It is powerfully built with shiny scales and stumpy limbs with 5 digits, although most adults lack some toes. Its name is derived from its shiny, fish-like head and body. It is a uniform dark brown to black on the back and sides, and adults have a conspicuous pale rim around the eye. Juveniles have bold cream or bluish-white spots on the flanks. It is white, yellow to orange-brown below. **Behaviour** A shy, ground-dwelling lizard, it is very alert and active during the day from spring to autumn, and hibernates in winter. They shelter in hollow logs, in the holes left by the roots of large fallen trees, or in extensive burrows excavated beneath logs or below thick ground cover. They live in family groups during the breeding season, often comprising more than one adult pair plus a number of associated juveniles. They often bask together in sunny spots on the forest floor or on logs, before heading off to forage independently. Individuals have discrete home ranges of some 10,000 square metres which they may share with other group members. They forage around dense vegetation, and can move very fast if disturbed. When startled or threatened they make snorting and hissing sounds, and dash noisily through the ground litter to the burrow. Outside the breeding season they are solitary. **Development** Females mate in early summer and bear 2–9 live young some 3 months later. Juveniles stay with their parents for some time before finding their own territory. They may live for up to 23 years. **Diet** Plant material such as fungi and fallen fruits make up a large part of its diet, together with insects, snails, slugs and mice. **Habitat** Rainforest margins and wet sclerophyll forests with canopy breaks close to dense understorey vegetation (including lantana and blackberry), and suburban gardens in coastal and near coastal areas.

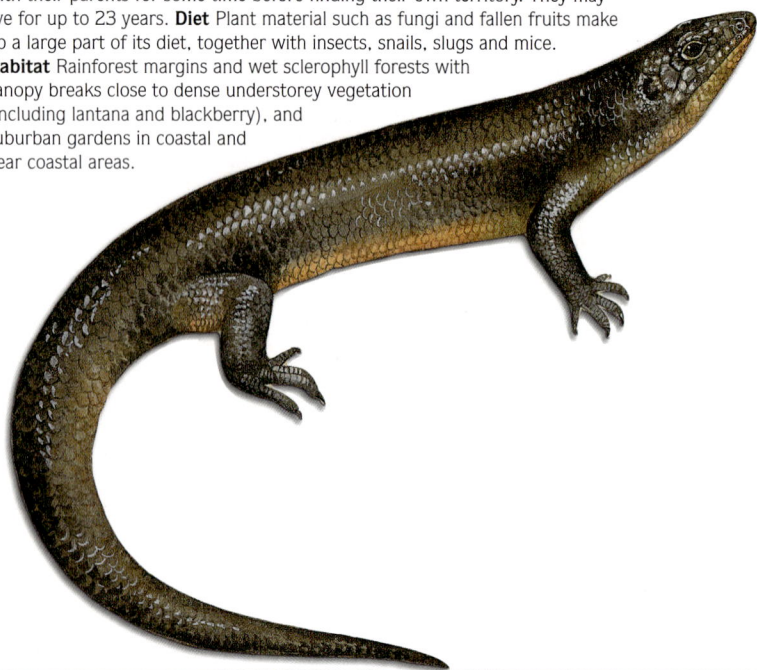

LENGTH **To 750 mm**   STATUS **Low risk.**

SPECIES *Egernia stokesii*

GIDGEE SKINK

A large, robust skink with a short head, a flattened body with short, backward-facing spines and a short tail with long spines. It is pale brown to dull reddish-brown above, sometimes with obscure dark bars over the back, and pale blotches on the sides of the head and neck. A large blackish-brown form occurs in the north and west, and darker individuals may have irregular rows of large whitish blotches across the body and tail. It is cream to pale brown below.

Behaviour This ground-dwelling, gregarious skink, is active by day and lives in colonies with up to 16 others in deep crevices in rocky outcrops, under large boulders and rock slabs. They shelter beneath surface debris such as logs and wood piles, and sometimes in split or hollow trees. Each colony occupies an exclusive territory marked by conspicuous piles of scats close to their shelter site, and colony members recognise each other by their chemical secretions. Colonies typically comprise breeding partners, their offspring and in some cases closely related adults. Breeding partners appear to be monogamous and maintain stable relationships over periods of at least 5 years. They bask close to the shelter and forage in the late afternoon and after sunset in hot weather, searching for prey in a home range within reach of the shelter site. If disturbed they dash to cover and if threatened in the shelter they make themselves very difficult to extract by inflating the body to wedge themselves in with their spiny scales gripping the sides. **Development** They mate in spring and early summer, and females bear a single litter of up to 8 live young in mid to late summer or early autumn. The newborn are about 108 mm long and reach sexual maturity at about 6 years. They have a potential lifespan of about 25 years. **Diet** Grasshoppers, crickets, moths, cockroaches and other arthropods, with some leaves, flowers and fruit. Adults consume more plant matter than juveniles. **Habitat** Arid and semi-arid acacia woodlands and shrublands, around rocky outcrops and outliers of the inland ranges. Also found on islands off the coast of WA.

LENGTH **To 285 mm**   STATUS **Low risk.**

FAMILY **Scincidae**	SPECIES *Egernia striolata*

Tree Skink

A medium-sized, fairly stout skink with a flattened head, body and tail, relatively smooth scales and well-developed limbs with 5 digits. It is pale grey, dark olive-grey or dark brown above, often scattered with small pale spots. The neck and shoulders are paler, and there is a dark band on the flanks between the eye and hips. The scales on top of the head usually have dark edges, and some individuals have narrow broken lines from the back of the neck to the tail. It is pale orange to dull yellow or silver-grey below, with dark bars on the chin and throat, and dark speckles on the chest. **Behaviour** This gregarious, secretive skink lives among trees or rocks and is active by day. In warm weather it shelters from the midday heat and forages in the early morning and late afternoon. In winter it is only occasionally seen on the surface basking in the sun. They are often found in a family group of adults and juveniles sheltering beneath the loose bark of living or dead trees, in splits and hollows in tree trunks and logs, in rock crevices or under rock slabs. Solitary individuals are sometimes found sheltering in small trees or logs. Groups have a regular defecation site near their nocturnal shelter, indicated by piles of scats. They bask close to the shelter and forage in the immediate surroundings, gleaning insects from beneath the bark and ambushing any smaller lizards that come close enough. They readily drop their tail if attacked and regenerate a new one. **Development** They mate in spring and females bear 2–6 live young in mid to late summer. The young are about 95 mm long at birth and become sexually mature at 2–3 years of age. **Diet** Mainly beetles, grasshoppers, ants, spiders, moths and other invertebrates, and occasionally smaller lizards. **Habitat** Rocky outcrops in dry sclerophyll forests and woodlands, with suitable crevices and rock piles.

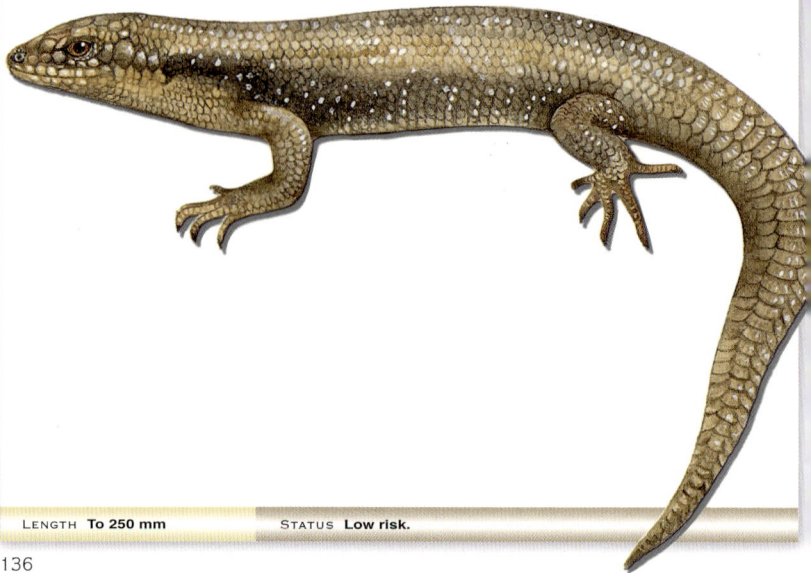

LENGTH **To 250 mm**	STATUS **Low risk.**

SPECIES *Liopholis whitii*

WHITE'S SKINK

A medium-sized skink with prominent creamy-white rings around its eyes, smooth scales and short limbs with 5 digits. It is pale to dark brown above, and almost black in alpine areas. Some individuals are uniform above, while most have 2 dark stripes running along the back, enclosing small white to cream spots. The flanks are usually paler with a few large, dark-edged pale spots, and a dark streak with a pale centre above the forearm. A pale stripe usually runs from the nostril, below the eye to the eardrum. It is pale grey to white below. **Behaviour** This swift-moving lizard is predominantly ground-dwelling, although it climbs around tree stumps, fallen timber and onto low rock faces to bask. It is active by day, and shelters at night in a shallow depression or a complex burrow system excavated beneath a large rock or log. It may also shelter in hollow logs and tree stumps, in rock crevices and in the abandoned burrows of larger animals. It is a gregarious species, living in small groups, usually comprising one or more adults and their juvenile offspring. They usually share the same burrow and tend to use a common basking area and defecation site close to the burrow entrance. Small piles of scats mark the shelter site and their home range. They forage on the ground close to the shelter site, and will sometimes jump up to catch a flying insect. If disturbed they dash into the burrow or under ground cover. In the late afternoon these lizards may climb as high as 600 mm in the vegetation to absorb the last warmth of the sun. **Development** They mate in spring and females give birth to 1–5 (usually 3) live young from late January to early March. They are about 85 mm long at birth, become sexually mature at the end of their second year, and may live for up to 8 years. **Diet** A wide variety of arthropods including beetles, ants, cockroaches, crickets and spiders, other invertebrates and plant matter including leaves, flowers and small fruits or berries. **Habitat** Dry sclerophyll forests, woodlands, tussock grasslands and heaths in alpine and lowland regions, mainly on slopes and in valleys. It is usually associated with rocks.

LENGTH **To 310 mm**   STATUS **Low risk.**

| FAMILY **SCINCIDAE** | SPECIES *Eremiascincus richardsonii* |

BROAD-BANDED SAND SWIMMER

A medium-sized skink with smooth, glossy scales, a pointed snout and well-developed limbs with 5 digits. It is pale yellow-brown to rich golden-brown above. Some individuals have no patterning, while others have 8–14 broad, dark cross-bands between the back of the neck and the hips, and 19–32 cross-bands on the tail. It is white or cream below, and many adults have regenerated tails. **Behaviour** This ground-dwelling lizard is active around dusk and at night. It shelters by day in a shallow burrow excavated in loose sand beneath leaf litter, below rotting timber or flat rocks, or in an abandoned burrow, often shared with 2–3 young in late summer. It frequently rests with its head protruding from sand or other cover, waiting to ambush passing prey. It also forages over relatively large distances. It is a very aggressive hunter and becomes agitated by the movements of its prey, waving its tail in a cat-like fashion before lunging at a victim. Large prey is grasped near the head, spun around to kill or break it up, and dragged back to the skink's shelter to be eaten. It escapes from predators by instantly burrowing into loose sand or soil with a wriggling, snake-like motion, hence the common name of sand swimmer. **Development** They mate in spring and females lay a clutch of 2–7 eggs, about 18 mm long, from October to February. They hatch in early to mid-autumn and the hatchlings are bright yellow with purple bands. They become sexually mature in their first year and may live to 10 years. **Diet** Arthropods, particularly moths, termites, beetles, grasshoppers and spiders. Rarely small skinks, geckos, snakes and mammals. **Habitat** Woodlands, shrublands and hummock grasslands on the plains and hills, with sand dunes or loose sand in depressions, from drier inland areas to sub-humid coastal sites.

| LENGTH **To 300 mm** | STATUS **Low risk.** |

SPECIES *Eulamprus quoyii*

EASTERN WATER SKINK

A medium sized, robust water skink with long limbs each with 5 digits, and a long tail slightly flattened from the side. It is metallic-brown to coppery or grey-brown above, usually with small black flecks or blotches. A sharply-defined, narrow, pale yellow stripe usually runs along each side from above the eye to the mid-body or base of the tail. The upper flanks are dark with small cream spots, while the lower sides are greyish-yellow, mottled with dark spots. The underside is white to yellow, often with dark lines running from the throat to the belly. **Behaviour** This fast-moving, predominantly ground-dwelling lizard is active by day and lives along the edges of creeks, lakes, swamps and other permanent waterways. In urban areas it sometimes lives in stormwater drains. It shelters in crevices among rocks or timber, in holes or burrows in the bank, probably made by other animals. It is often seen basking in the sun on a log or rock by a waterway, or foraging among waterside vegetation. Large individuals are sometimes seen on rocky ridges well away from water. Inland populations forage around the root systems of large riverside eucalypts, while in coastal areas they often forage in intertidal zones where fresh water seeps out among rocks. It can usually be approached quite closely, but if startled it dashes to cover, or plunges into the water and swims quickly to the opposite bank where it disappears into its shelter. It is strongly territorial and will fight savagely to keep other water skinks out of its home range. In cool areas it becomes inactive over winter and emerges from its shelter site in spring. **Development** They mate in spring and females give birth to 2–9 live young in summer. They become sexually mature in their first year and may live to 10 years. **Diet** A wide variety of invertebrates including worms, water beetles and aquatic larvae. Also tadpoles, small frogs, fish, snails, smaller lizards and berries. **Habitat** Rainforests, wet and dry sclerophyll forests, woodlands and heaths in the ranges, slopes and lowlands, particularly in rocky areas around the edges of creeks, rivers and swamps, and other moist sites. Also found around the Darling and Murray drainage systems in the semi-arid interior.

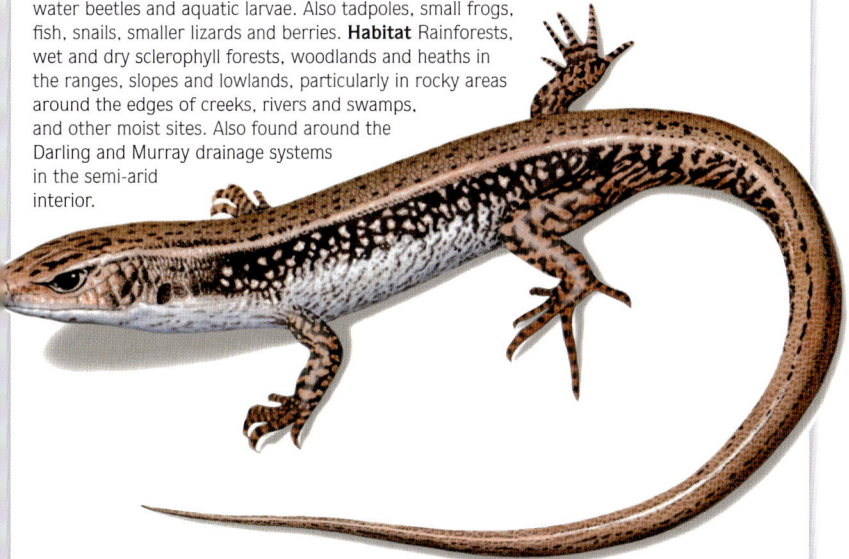

| LENGTH **To 300 mm** | STATUS **Low risk.** |

| FAMILY **SCINCIDAE** | SPECIES *Eulamprus tenuis* |

BAR-SIDED OR RAINFOREST SKINK

A medium-sized skink with long, slender limbs each with 5 well-clawed digits. It is smooth above, pale fawn to rich brown or silvery-brown, with small dark blotches scattered or aligned in bars, and sometimes forming a narrow stripe along the spine. A dark stripe extends from the side of the head to the hips. The lower flanks are pale grey spotted with black. The tail is marked with narrow black bands or blotches. The underside is white to pale yellow with grey marks on the throat. **Behaviour** This agile, secretive skink lives mainly in trees and sometimes around rocks. It is active during the day and usually forages in the early morning and around dusk, climbing among trees and rocks searching for insects. It prefers moist areas, sheltering beneath loose bark, in tree hollows up to about 1.5 m above ground. In some areas it shelters in rock crevices, beneath flaking rock and in caves. It is encountered in urban areas where it hides in stone walls. Small groups of adults and juveniles are sometimes found together, suggesting a social structure. It may be seen basking in the sun in the early morning, but is very shy and stays close to its shelter, retreating to cover at the slightest disturbance. **Development** Mating probably takes place in spring and females bear 2–7 live young in mid to late summer. **Diet** Spiders, ants, beetles, grasshoppers, insect larvae and pupae, other invertebrates and berries. **Habitat** Rainforests and vine thickets, wet sclerophyll forests, moist refuges in dry sclerophyll forests and woodlands along the coast and ranges. It is also found in gardens.

| LENGTH **To 200 mm** | STATUS **Low risk.** |

FAMILY **SCINCIDAE**   SPECIES *Eulamprus tympanum*

SOUTHERN WATER SKINK

A medium-sized, robust water skink with long limbs and a slightly flattened tail. It is golden-brown to dark brown or almost black above with irregular black blotches. A pale stripe runs from the lower lip to the ear. The upper sides are black, scattered with small white to yellow spots. The lower sides are pale grey spotted with white and black. The chest and belly are cream to yellow with black flecks, and the throat is white to pale grey. **Behaviour** This ground-dwelling to semi-arboreal skink is one of the most commonly encountered reptiles in montane forests, where it becomes torpid in the winter months. A sun-loving species, it is active by day, and frequently basks on rocks or logs. It is often found in wetter forests well away from water, but in lower rainfall areas it lives along the margins of watercourses. It shelters in a burrow system dug beneath low vegetation, around tree roots, under logs or rocks, in rocky crevices, dry stone walls or in rotting timber. Most of its activity is centred on rotting logs and stumps which are used as perches for basking and cavities for shelter both at night and over winter. If alarmed it flees to its shelter or drops into the water and swims below the surface to the other bank. It will drop part or all of its tail to escape a predator. Adults occupy fixed home ranges and vigorously attack intruders. **Development** They mate in spring and females bear 1–6 live young, about 75 mm long, in mid to late summer. Females may live to 13 years and males to 11 years. **Diet** Invertebrates including insects, spiders and snails. Also tadpoles, small frogs, lizards and a small amount of plant matter. **Habitat** Open forests, woodlands, shrublands and tussock grasslands, around creeks, rivers, swamps and well-watered slopes, from the highlands to cool lowland sites.

LENGTH **To 250 mm**   STATUS **Low risk.**

FAMILY **SCINCIDAE** SPECIES *Cyclodomorphus gerrardii*

PINK-TONGUED SKINK

A large skink with a relatively slender body and a large, squarish head. It has a long, slender, prehensile tail and short limbs with strongly-clawed digits. It is pale to pinkish-brown or silver-grey above, with 5–8 broad dark bands running obliquely across the back, and 8–12 black bands across the tail. Juveniles are very brightly marked, while some older adults have no bands. The underside is white to dark pinkish-brown, sometimes with dark marbling or broad bands. Pink-tongued skinks are born with a blue tongue which usually changes to pink when they are 1–2 years old. **Behaviour** This secretive skink is essentially ground-dwelling, although it often climbs into low vegetation using its prehensile tail as a fifth limb. It is active at night and around dawn and dusk, and often basks in the sun in the upper branches of shrubs in forest clearings. In winter it can be seen during the day. It shelters in tree hollows and cracks in the bark, in hollow logs, beneath deep leaf litter and in rock crevices. It is most active on warm, wet nights, when it forages for slugs and snails, which it crushes using a very large rounded tooth at the back of its jaw. In dense undergrowth it pulls itself along using only its front legs, holding its hind legs close to its sides. If threatened it raises its fore-body off the ground and flicks its tongue in and out in a snake-like fashion. It is quite aggressive and can inflict a painful bite with its strong jaws. **Development** They mate in spring, and females give birth to a litter of 5–33 live young, 70–110 mm long, in summer. **Diet** Mainly snails and slugs, supplemented by spiders, crickets, cockroaches and other small arthropods. **Habitat** Rainforests, wet sclerophyll forests and damp woodlands along the coastal plains and eastern slopes. Also found in well-watered suburban gardens in Brisbane.

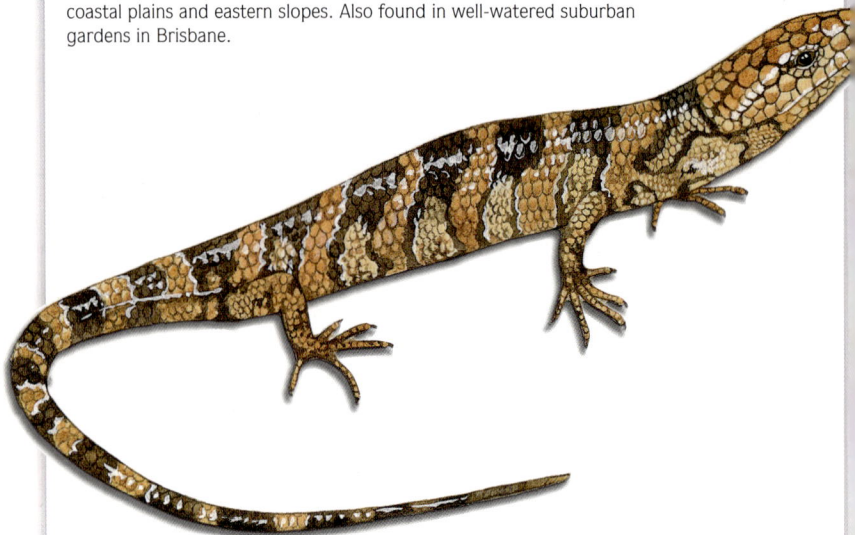

LENGTH **To 450 mm** STATUS **Low risk.**

FAMILY **SCINCIDAE**   SPECIES *Lampropholis delicata*

DARK-FLECKED GARDEN SUNSKINK
GRASS SKINK

A small skink with a fragile tail, often found in gardens in eastern Australia. It is coppery-brown to brownish-grey above, sometimes with scattered dark and pale flecks or streaks, occasionally forming short bands across the tail. The upper flanks are usually darker, and some females have a pale narrow line along the flanks. The undersides are white to cream, often with dark flecks or lines on the throat. **Behaviour** This common skink is mainly ground-dwelling, although it occasionally climbs among the lower branches of small trees, onto garden fences and walls. It is active by day and on warm summer evenings. It is usually seen basking on logs, leaves and pathways, and foraging among low vegetation and leaf litter. It shelters at night among leaf litter, below fallen timber, under logs and stones. In bushfires it survives by hiding in the deep burrows of spiders and insects and in cracks in the soil. Large populations often occur in suburban gardens where they occasionally fight and chase each other. They sometimes engage in a ritual display, moving their back legs in a slow walking motion. When attacked they readily shed their tail and regenerate a new one. **Development** In spring and early summer, females lay 1–8 eggs, about 8 mm long, in a rock crevice, in a hole in the ground or under a log or rock. Communal nesting sites are often used, with more than 400 eggs being laid in one site. The eggs hatch 1–2 months later in mid to late summer. The hatchlings are about 38 mm long. In a good season females may lay a second clutch in mid-summer. **Diet** Spiders, ants, beetles, bugs, centipedes, flies, insect larvae and other arthropods. **Habitat** Cool temperate to tropical rainforests, wet and dry sclerophyll forests, woodlands, grasslands and coastal heaths. Commonly found in suburban gardens.

LENGTH **To 105 mm**   STATUS **Low risk.**

COMMON GARDEN SUNSKINK

A small, fairly robust skink with a coppery-brown head and neck and a fragile tail. It is grey, olive-brown or coppery-brown above, scattered with small dark flecks and larger pale flecks, sometimes aligned into bars across the tail. A dark stripe usually runs along the spine from the head to the base of the tail. The upper sides often have a dark brown stripe extending from the nostril through the eye to the hips or tail, with a pale line below. The undersides are white to grey, sometimes scattered with dark markings. **Behaviour** This skink is one of the most abundant skinks in suburban gardens. It is active by day and is usually seen basking or foraging for small insects among leaf litter, grass, rock piles and low vegetation. It shelters under rocks, logs and ground debris, frequently sharing its shelter and basking sites. It is nomadic and moves to a new area every few weeks. Local populations can become very dense in good habitats. In these conditions large clumps sometimes form in the spring mating season, with up to 18 entwined individuals forming a ball of skinks each grasping another with its mouth. They may stay entwined for up to 4 minutes. Dominance hierarchies are also established by biting and neck arching, and large males usually dominate the groups. When attacked they readily shed their tail and regenerate a new one. **Development** They mate in late winter and spring and females lay a clutch of 1–4 eggs, about 9 mm long, in a communal egg-laying site where more than 200 eggs may be laid by different females. They hatch from late summer to early autumn and mature in 8–9 months. More than one clutch may be laid in a good season. **Diet** Small insects including flies, ants and moths, worms, and a variety of arthropods. **Habitat** Dry sclerophyll forests, open woodlands and coastal heaths. It is sometimes found at the edge of rainforests, in wet sclerophyll forests and moist tussock grasslands, and is common in suburban gardens.

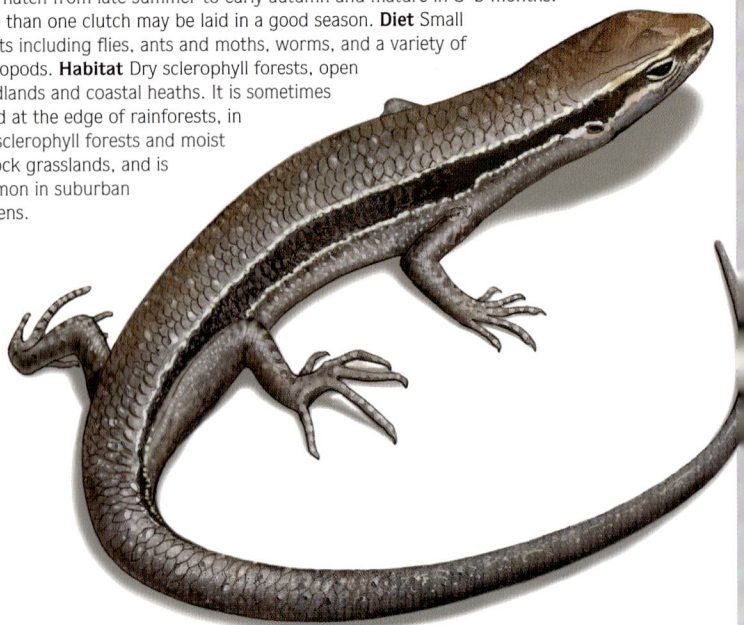

LENGTH **To 100 mm**	STATUS **Low risk.**

BOUGAINVILLE'S SKINK
SOUTH-EASTERN SLIDER

A small, slender, elongated skink with a fragile tail and short, thin limbs, each with 5 digits. The forelimbs are about half the length of the hind limbs. It is pale brown to greyish-brown above, with 4 thin lines or dashes along the back and a broad dark stripe extending from the snout through the eye to the base of the tail. The lower sides are white to grey, streaked with black, and in some populations the tail is flushed with brown, yellow or red. The underside is white with yellow or red below the tail. Northern individuals often have black spots on the chest and belly. **Behaviour** This ground-dwelling skink is usually only seen at night, although it is active under cover during the day, burrowing into loose soil beneath logs, rocks and leaf litter, foraging for insects and ambushing any prey that comes close to its resting site. At night it emerges to feed on the surface, and if disturbed it plunges into the leaf litter or into loose, sandy soil, wriggling its body into the sand in a 'sliding', sideways motion, hence its common name. This behaviour leaves the distinctive wavy tracks that often criss-cross sandy areas. **Development** On the mainland they mate in spring and females lay 2–4 soft-shelled eggs in summer beneath surface cover. The eggs hatch 1–2 months later. In Tasmania, Kangaroo Island and the Bass Strait Islands they mate in autumn and females store the sperm over winter, giving birth from mid to late February to 2–4 fully developed young enclosed in transparent membranes. They emerge from the membrane a few days later. **Diet** Termites, ants, other insects and their larvae, and spiders. **Habitat** Sclerophyll forests and woodlands, open heathlands, shrublands and rocky outcrops on the slopes, ranges and coastal plains.

LENGTH **To 125 mm**   STATUS **Low risk.**

FAMILY **SCINCIDAE**   SPECIES *Morethia ruficauda*

LINED FIRE-TAILED SKINK

A small, strikingly-coloured, elongated skink with a long head, a fragile tail and well-developed limbs with 5 digits. It is smooth and glossy black above with 2 prominent, broad, cream to white stripes along the sides from the snout to the base of the tail. The lower back and hips are flushed with reddish-orange, which increases in intensity on the tail. The limbs are also flushed with red. The underside is white.
Behaviour This ground-dwelling skink is active by day, often during high midday temperatures. It is swift and agile, and dives into leaf litter or runs to cover if disturbed. It shelters among rocks and leaf litter, and basks and forages around the base of large boulders. Its colourful tail is used as a means of communication, probably related to territorial or mating behaviour. Individuals frequently curl their tail and slowly wave it over their head and from side-to-side. They also face each other at a distance of about 100 mm and whip their tails from side-to-side. **Development** Females lay 1–3 eggs and the hatchlings have bright red tails. **Diet** Spiders, ants, beetles, moths and other arthropods. **Habitat** Tropical woodlands, shrublands, hummock and spinifex grasslands in sub-humid to arid areas with rocky outcrops and hills on sandy, clay or stony soils.

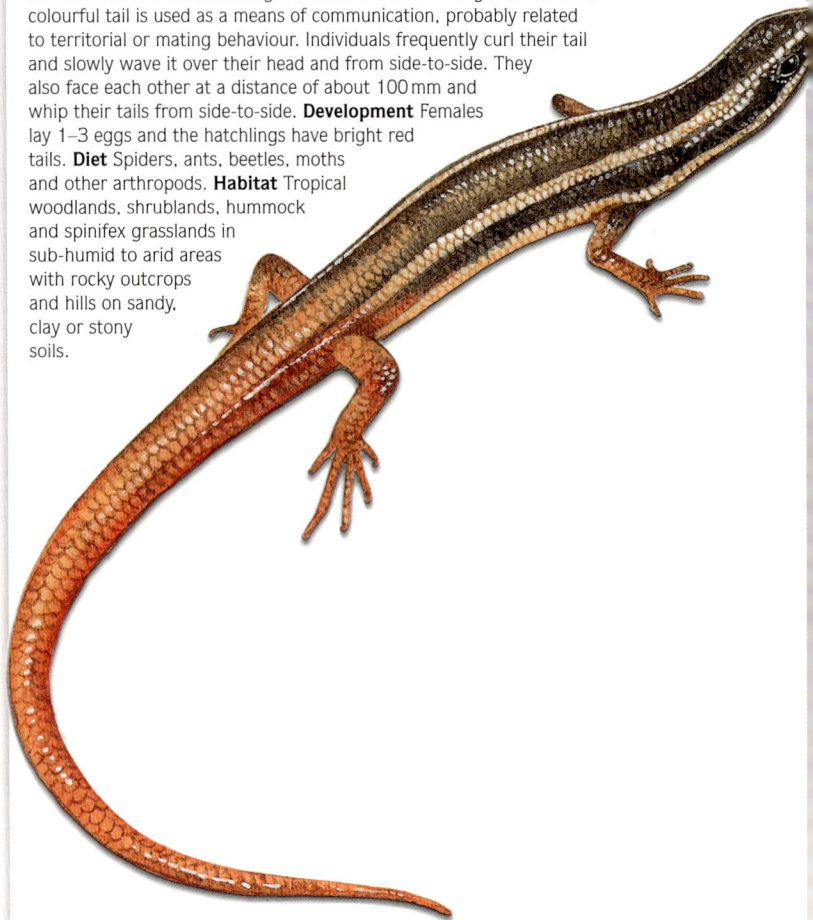

LENGTH **To 90 mm**   STATUS **Low risk.**

FAMILY **SCINCIDAE**   SPECIES *Saiphos equalis*

THREE-TOED SKINK

This small burrowing skink is sometimes mistaken for a small snake. It has very short limbs, each with 3 small toes, a smooth, elongated body and small eyes. It is shiny coppery-brown to grey-brown above with fine black longitudinal lines and a black tip to the tail. The flanks and tail are blackish-brown and the underside is yellow to bright yellowish-orange. Some individuals have black throats. **Behaviour** This ground-dwelling and burrowing skink is active at night and around dawn and dusk, and is rarely seen on the surface in the open. It shelters and forages among leaf litter, under rocks and logs in moist sites. In friable and sandy soils it wriggles its way into the soil to hide, and in clay soils it digs permanent tunnels. It also sheds its skin underground. This skink is frequently found in large numbers in gardens and compost heaps. It thrashes around violently when disturbed and attempts to burrow back into the ground. When moving slowly it uses its limbs, but when moving fast it tucks its limbs alongside its body and moves like a snake with lateral undulations. **Development** The mode of reproduction of this skink varies according to the location. In the lowlands the females lay eggs with varying amounts of calcium in the shells. Along the south coast they take 1–9 days to hatch, while on the north coast they take up to 30 days to hatch. At elevations over 1000 m, the young are fully developed and surrounded by a clear membrane which they break out of in a few hours. Females reproduce in late summer. Litter sizes range from 1–7, and the hatchlings are about 55 mm long. **Diet** Crawling insects and larvae, worms, centipedes and other arthropods. **Habitat** Rainforests, wet and dry sclerophyll forests, woodlands, subalpine tussock grasslands, heathlands and suburban gardens along the coastal lowlands and adjacent ranges.

LENGTH **To 200 mm**   STATUS **Low risk.**

FAMILY **SCINCIDAE**   SPECIES *Saproscincus mustelinus*

WEASEL SKINK

A medium-sized skink with a long fragile tail and long limbs each with 5 digits. It is brownish-grey, orange-brown to dark brown above usually with scattered flecks. It often has paler flanks, with scattered pale and dark flecks sometimes forming lines. An orange-brown stripe runs along the side of the tail and there is a conspicuous white or cream bar behind the eye. The underside is white to bright lemon-yellow with dark streaks from the throat to the tail. **Behaviour** This ground-dwelling skink is active mainly around dawn and dusk and during the day in dull overcast weather. It tends to avoid direct sunlight and shelters beneath leaf litter, rotting wood or rocks in damp sites where there is dense ground cover. It is also found in suburban gardens close to rainforests and moist gullies. It is sometimes seen basking on leaves or logs in cool conditions, but it is very secretive and is rarely seen foraging. In winter it becomes inactive and hibernates alone or in small aggregations of up to 21 individuals under rotting logs or dense ground cover. If attacked this skink will readily drop its tail and regenerate a new one. **Development** Females lay 1–2 clutches of 2–9 eggs about 9 mm long, from late spring to late summer. The second clutch is smaller than the first. Several females often use the same egg-laying site, usually under ground litter or a rock. Communal sites may contain more than 136 eggs. The eggs hatch about 1 month later. The hatchlings are about 50 mm long. **Diet** Mainly small insects, insect larvae and other invertebrates. **Habitat** Wet sclerophyll forests, temperate rainforests, woodlands and heathlands along the coast to the adjacent ranges. Also in suburban gardens and degraded areas such as grazing lands.

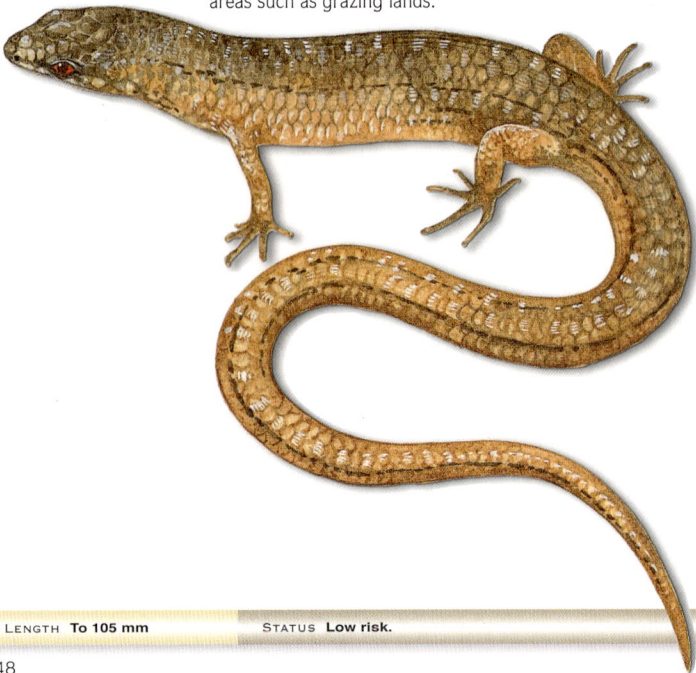

LENGTH **To 105 mm**   STATUS **Low risk.**

CENTRALIAN BLUE-TONGUED LIZARD

A large, robust skink with a narrow neck and a broad, angular head with a wide, fleshy, bright blue tongue. It has a short body and limbs, a short, slender tail and smooth scales. It is pale grey to greyish-brown above with 9–13 narrow orange to yellow-brown cross-bands from the back of the neck to the hips, and another 8–10 wider, darker, cross-bands on the tail. There is a conspicuous broad black stripe from the eye to above the eardrum, and the upper parts of the limbs are usually black. The underside is white to cream, often with dark marks on the throat. **Behaviour** This slow-moving, solitary, ground-dwelling skink can be active at any time of the day or night, although it is usually seen in the early morning, late afternoon and on warm nights. In cool weather it is often seen crossing roads during the day. It shelters in the middle of grass hummocks, in the abandoned burrows of other animals, in diggings under stones, and in leaf litter. When alarmed it gapes widely, displaying the bright pink lining of its mouth and its large, blue, protruding tongue. At the same time it inflates and deflates its body and hisses threateningly. **Development** They mate in spring and females give birth to 2–10 live young in summer. The newborn are about 120 mm long. **Diet** Insects and other arthropods including grasshoppers, beetles and bugs. Also carrion, flowers, fruit and other soft plant matter. **Habitat** Open woodlands, open shrublands and hummock grasslands with scattered shrubs and trees, in semi-arid to arid red sand plains, dunes and stony hills. **Threats** Degradation of habitat by grazing stock, predation by cats and foxes.

LENGTH **To 450 mm**   STATUS **Vulnerable in NSW.**

| FAMILY **SCINCIDAE** | SPECIES *Tiliqua nigrolutea* |

BLOTCHED BLUE-TONGUED LIZARD

A large, robust skink with a narrow neck and a deep, short head with a wide, bright blue tongue. It has a long body, smooth scales and a short, thick tail. The small limbs each have 5 digits. It is dark brown to black above with large, oblong, yellowish, pale brown or orange blotches, scattered or aligned in longitudinal rows or cross-bands on the body, and forming bands around the tail. The flanks have larger and paler blotches. The top of the head is olive-brown and the side of the head is flushed with yellow or orange. The undersides are cream to yellow. Southern animals are smaller and less brightly coloured than the NSW alpine form. **Behaviour** This slow moving, ground-dwelling skink is active by day and at low temperatures. It is a sun-loving lizard and is often seen basking on roads. It shelters at night in hollow logs, in the abandoned burrows of other animals, in diggings beneath stones and in deep leaf litter. Like the other blue-tongues, it opens its mouth wide when alarmed and thrusts out its bright blue tongue which contrasts starkly with the pink mouth-lining. At the same time it hisses loudly while inflating and deflating its body. It has powerful jaws and can inflict a crushing bite. **Development** Mating takes place in spring and females give birth to 2–15 large, live young, about 150 mm long, in late summer or early autumn. Females in southern areas do not usually breed every year. They have a lifespan of 20 years. **Diet** Insects, snails, slugs, worms, spiders, mice, carrion, fungi, flowers, leaves and native berries. **Habitat** Tall open forests, woodlands, heathlands, alpine meadows and swamps, from the coast to the ranges. Also occurs in suburban gardens.

| LENGTH **To 500 mm** | STATUS **Low risk.** |

FAMILY **Scincidae**   Species *Tiliqua occipitalis*

Western Blue-tongued Lizard

A large, robust skink with a relatively small, flattened head with a broad, bright blue tongue. It has short limbs with 5 digits, a short tail and smooth scales. It is pale yellowish-brown above with 4–7 broad dark brown bands on the body and 3–4 bands on the tail. Pale spots or large open circles may be found within the bands on the back. A broad, dark, glossy stripe runs from the eye to the eardrum. The undersides are cream to white. **Behaviour** This slow moving ground-dwelling lizard is active by day. It usually shelters in a rabbit warren or other abandoned burrow, beneath deep leaf litter, under rocks or fallen timber. When harassed it opens its mouth wide to display its large blue tongue and hisses threateningly while puffing up its body. **Development** They mate in spring and females give birth to 3–10 large, live young, about 160 mm long, in summer. They have a lifespan of up to 25 years. **Diet** Insects, snails, carrion, leaves, flowers, native berries and fungi. **Habitat** Woodlands, shrublands, mallee, heathlands, hummock and spinifex grasslands from sub-humid to arid areas. Also among red sand ridges in the NT, and in coastal sand dunes. **Threats** Land clearing and degradation of habitat by grazing stock, predation by cats and foxes.

LENGTH **To 400 mm**   STATUS **Vulnerable in NSW. Near threatened in Vic.**

| FAMILY **SCINCIDAE** | SPECIES *Tiliqua rugosa* |

SHINGLEBACK LIZARD

A large, very robust skink with a large, broad, triangular head and a wide, greyish-blue tongue. It has large rough scales with a dull shine and a short, blunt tail used to store fat. The short, strong limbs each have 5 digits. It is yellowish-brown to black above, usually marked with 5–7 narrow and indistinct to broad, pale yellow, orange or grey bands on the body and tail. The head is usually paler and is often flushed with orange. The undersides are cream to white usually with grey to brown stripes, bands or blotches. **Behaviour** This slow-moving, ground-dwelling lizard is active by day and is often seen basking on roads. It shelters in hollow tree trunks, abandoned burrows, beneath fallen timber, in deep leaf litter and under low vegetation. They are solitary and live in overlapping home ranges of around 4000 square metres for most of the year, staying in the same home range for several years. Rather than having a central refuge, they shelter under cover anywhere in the home range. Pairs form in early September and stay together for 6–8 weeks, spending most of their time in contact or less than 500 mm apart. Males defend their female partner from rival males, until they separate after mating. They are monogamous, and partners re-form over successive years. If cornered it inflates its body, opens its mouth wide to display its blue tongue and hisses loudly. It is generally inoffensive, but can inflict a painful bite if provoked. **Development** They mate in spring, and females give birth to 1–3 large young, about 150 mm long, in late summer or early autumn, after about 5 months gestation. They become sexually mature in 3–4 years and may live to 20 years or more. **Diet** Mostly ground blossoms and green plant matter, also insects, snails and carrion. **Habitat** Dry sclerophyll forests and woodlands, to hummock grasslands and shrublands, along the coast and hinterland, in arid and semi-arid areas.

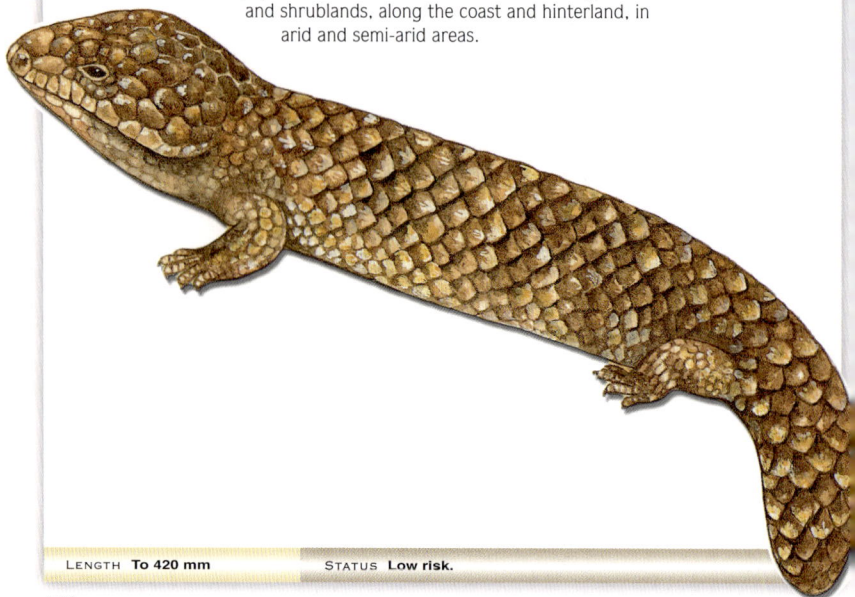

| LENGTH **To 420 mm** | STATUS **Low risk.** |

FAMILY **SCINCIDAE**   SPECIES *Tiliqua scincoides*

EASTERN BLUE-TONGUED LIZARD

A very large, robust skink with a slightly flattened head and a broad, bright blue tongue. It has short limbs with 5 digits, a short, thick tail and smooth scales. The colour varies above through shades of yellow, brown, pale grey to black, patterned with 6–9 dark-edged, paler bands on the back and another 7–10 bands on the tail. A broad dark stripe usually runs from the eye to the eardrum. The underside is white to grey or yellow, often with pale brown marks. Northern individuals are slightly larger with indistinct bands on the back, often broken up into fine mottling. **Behaviour** This relatively slow-moving, generally shy skink is most active in the early morning and late afternoon, and also on warm nights, and often frequents gardens and parks. It is a sun-loving lizard and is usually seen basking on roads. It shelters in abandoned burrows, beneath fallen logs, in crevices and hollows. It retreats at any sign of a threat, and if cornered it attempts to frighten its aggressor by opening its mouth wide, inflating its body and sticking out its bright blue tongue while hissing loudly. As a last resort it may inflict a painful though superficial bite if mishandled. When crawling through dense vegetation it tends to drag its hind legs and is sometimes mistaken for a death adder. **Development** They mate in late winter and spring, and females give birth to 5–25 live young from early to late summer. The newborn are about 130 mm long and they may live to 20 years. **Diet** Snails, mice, carrion, insects, flowers, leaves, fruit and fungi. **Habitat** Most habitats excluding rainforests, from coastal heaths, woodlands and forests to montane forests and drier woodlands and grasslands of the inland plains. Often found in suburban gardens.

LENGTH **To 600 mm**   STATUS **Low risk.**

FAMILY **PYGOPODIDAE**	SPECIES *Delma tincta*

EXCITABLE DELMA

This medium-sized legless lizard can be distinguished from a snake by its broad, unforked fleshy tongue and conspicuous ear openings in the sides of its head. It has snake-like, lidless eyes, lacks forelimbs and its hind limbs are reduced to 2 small, scaly flaps near the vent. It is smooth and shiny, grey to grey-brown above and white below. A number of wide dark bands usually cross the head, separated by narrow pale cream or yellow bands. **Behaviour** This legless lizard is active on the surface during the day and early evening and usually moves with a snake-like motion, but in open areas it propels itself forward in a series of rapid leaps. It hides in loose soil and leaf litter, beneath rocks, under logs, in soil cracks, in abandoned termites' nests and in hummock grasses. It is also able to burrow, using its snout in a sideways motion to dig into the soil. When sheltering it usually leaves its head exposed, ready to ambush passing prey. It has a fragile tail which it readily discards as a sacrifice when attacked by a predator. If grasped it makes loud wheezing squeaks. When disturbed it becomes very agitated, leaps from the ground, flicking itself up and forward with its tail, and contorting its body in the air. This makes the lizard very difficult to catch. **Development** Females lay a clutch of 1–2 elongated parchment-shelled eggs, about 13 mm long, in early September. **Diet** Insects, spiders and other arthropods. **Habitat** Open woodlands, shrublands, hummock grasslands and tree-lined arid-zone watercourses, with rocky rises and sandy soils. Favours open areas with a grassy understorey.

LENGTH **To 267 mm**	MIDBODY SCALES **12–16 rows**	STATUS **Low risk.**

FAMILY **PYGOPODIDAE**   SPECIES *Lialis burtonis*

BURTON'S LEGLESS LIZARD

A snake-like lizard with a robust body and an elongated head with a long, sharp, wedge-shaped snout and small eyes with vertical pupils. It can be distinguished from a snake by its broad, fleshy tongue, conspicuous ear openings and relatively short tail. It lacks forelimbs and has small, scaly, triangular flaps on either side of the vent instead of hind limbs. It has smooth scales, varying greatly in colour from pale grey or cream, through shades of brown to almost black. It may be plain or patterned with stripes or blotches. There is often a cream or white stripe along the side of the head and neck, and the underside is covered with small dark and pale flecks. **Behaviour** This ground-dwelling lizard is active during the day and at night, but is most frequently encountered moving around in the first hours after dusk and in the early morning. It shelters below rocks, under logs, among leaf litter and in hummock grasses and abandoned burrows. It moves with a snake-like action, stalking and ambushing its prey, often moving almost undetectably slowly to get into striking range. It often lies in wait under cover or in a clump of vegetation until a small lizard comes close enough to catch, sometimes raising and shaking its tail as a lure. It pounces on its victims with great speed and accuracy, biting and holding them until they are subdued. It sometimes climbs into low, dense vegetation to forage, and often stays in its shelter for days without moving. If attacked it will readily drop its tail as a sacrifice and grow a new one. **Development** They mate in late winter and early spring in southern areas, and females lay a clutch of 1–3 parchment-shelled eggs, about 22 mm long, in summer, probably producing more than one clutch in a season. The eggs are laid under logs or rocks, or on the soil, and sometimes in the nests of sugar ants. Communal nests may be used, and 20 eggs have been found in one nest. The hatchlings are about 136 mm long. They may mate at other times of year in the north. **Diet** Small reptiles, including skinks, legless lizards, geckos and dragons, and some invertebrates. **Habitat** Spinifex grasslands, beaches, tall open heaths and woodlands, from deserts to the margins of rainforests.

LENGTH **To 645 mm**   MIDBODY SCALES **18–22 rows**   STATUS **Low risk.**

| FAMILY **PYGOPODIDAE** | SPECIES *Pygopus lepidopodus* |

COMMON SCALY FOOT

The largest of the legless lizards, it is robust and blunt-nosed, and can be distinguished from a snake by its broad, fleshy tongue and conspicuous ear openings. It lacks forelimbs and has relatively large, conspicuous, paddle-shaped flaps instead of hind limbs. The scales are matt above and glossy below. Its colour varies from uniform grey to reddish-brown above, sometimes with a pattern of black dashes or blotches in rows down the sides of the body and tail. The lips and side of the neck usually have a few dark bars, and eastern individuals may have a contrasting grey head and tail. Females are larger than males. **Behaviour** This ground-dwelling lizard is mainly active on warm, sunny mornings from spring to autumn, but is active at night on hot days. In cold conditions it hibernates. It forages in open areas or among low vegetation, climbing among dense shrubs and hummock grasses, and looks for spider burrows around the edges of low shrubs. It shelters below low vegetation, under fallen timber, beneath surface debris and leaf litter and in dense grasses and shrubs, and has been observed burrowing. Adults sometimes flick themselves up on their tails and move by side-winding. If threatened it mimics the actions of a poisonous snake, raising its head and fore-body high off the ground, flattening its neck and flicking its fleshy tongue in and out, sometimes striking out with the mouth closed. If caught it rotates its body violently and emits a loud wheezing squeak.
Development They mate in spring, and females lay a clutch of 2 parchment-shelled eggs, about 40 mm long, in summer, and may lay more than one clutch per season. They use communal egg-laying sites beneath a log or rock, and 76 eggs have been found in one nest. The eggs hatch about 106 days later, from March to late April, and the hatchlings are about 210 mm long.
Diet Spiders, scorpions, insects, frogs and some native fruits. It laps up the body fluids of mutilated spiders and scorpions, crushes spider egg sacs and licks rainwater from vegetation.
Habitat A wide variety of habitats including dry sclerophyll forests, the margins of wet sclerophyll forests, semi-arid mallee woodlands, coastal heaths and sand dunes behind beaches. Found above 1000 m in the Gibraltar Ranges.

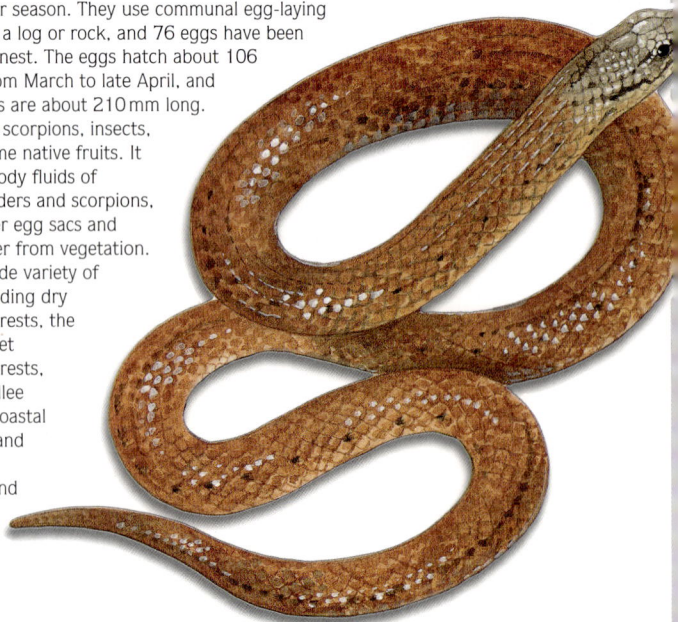

| LENGTH **To 840 mm** | MIDBODY SCALES **21–25 rows** | STATUS **Low risk.** |

FAMILY **PYGOPODIDAE**   SPECIES *Pygopus nigriceps*

WESTERN HOODED SCALY FOOT

A medium-sized, robust, legless lizard, it can be distinguished from a snake by its broad, fleshy tongue and conspicuous ear openings. It lacks forelimbs and has noticeable paddle-shaped flaps instead of hind limbs. It is smooth and glossy, pale grey to reddish-brown above, sometimes with dark-edged scales forming a net-like or vague longitudinal pattern. The head and neck have a dark hood comprising a narrow band across the eyes and a wider collar across the back of the neck. This may merge into a single band in older individuals or become very faint, blending into the background colour. It is white below. **Behaviour** This legless lizard is ground-dwelling and is generally active on the surface at night and around dawn and dusk, although in cool conditions it forages by day, and is probably inactive in winter. It hunts on the surface and in soil cracks and burrows. Scorpions and other large arthropods are subdued and crushed by twisting and rolling, similar to a crocodile-roll. Their body fluids are then licked up. It shelters beneath rocks, under fallen timber, below surface debris, in hummock grasses, in soil cracks, in abandoned burrows and termite mounds. When disturbed it may mimic the actions of a poisonous snake, raising its head and fore-body off the ground, flattening its neck and occasionally striking out. This bluff is often enough to deter a potential predator. It readily sacrifices its tail if attacked and regenerates a new one. **Development** They mate in spring and early summer and females lay a single clutch of 2 parchment-shelled eggs, about 18 mm long. Hatchlings emerge after an incubation period of some 73 days and are about 176 mm long. **Diet** A variety of arthropods, mostly insects with some spiders, scorpions, and spider egg sacs. **Habitat** Found in a variety of habitats including hummock grasslands, open woodlands and shrublands, from red sand ridges to black-soil plains, in sub-humid to arid areas. **Threats** Habitat destruction, fragmentation and degradation due to land clearing, grazing, logging and firewood collection, predation by feral foxes and cats.

LENGTH **To 575 mm**   MIDBODY SCALES **21–25 rows**   STATUS **Critically endangered in Vic.**

FAMILY **TYPHLOPIDAE**   SPECIES *Ramphotyphlops nigrescens*

COMMON EASTERN BLIND SNAKE

NON-VENOMOUS, HARMLESS

The largest Australian blind snake, this worm-like reptile has a blunt head indistinct from its neck and tiny, rudimentary eyes. It has a very short tail with a small, hooked spine at the end, used as an anchor or a digging spike to push through hard soil. The mouth is small and well back from the rounded snout. The scales are small, tough and shiny, and overlap to reduce friction while digging and to protect the snake from ant bites. It is purplish-brown to almost black above with pale edges at the base of the scales, creating a vague net-like pattern. There is usually a dark patch on each side of the anal region, and the underside is cream to pinkish. Males are smaller than females. **Behaviour** This secretive, burrowing snake is usually only seen on the surface at night after rain, floods or in warm, humid weather. Although active by day it shelters under rocks, logs, in moist soil beneath leaf litter, in abandoned termite nests and rotten tree stumps up to 1 m above ground. It can follow ant trails up to a week old, pushing its way through soft soil into the nest, flicking its pale, forked tongue in and out to detect their odour. Glands beneath the head and neck scales secrete a smelly, oily fluid that probably lubricates the scales, helping it to slip through small spaces. The tiny eyes can detect light and dark but little else. It lacks an ear aperture, but responds to vibrations. If disturbed it often curls up into a loose ball with its head hidden and its tail sticking up to display its dark anal markings. If alarmed it gives off a pungent odour from its anal glands, which is not enough to deter large ants whose stings are able to kill small blind snakes. Writhing knots of up to 35 individuals of both sexes, presumably mating, have been found under surface cover in the breeding season. **Development** They mate in late winter and spring, and females lay a clutch of 5–20 soft-shelled eggs, about 14 mm long, in mid to late summer. Eggs have been found on soil beneath a rock. Hatchlings, about 120 mm long, emerge 30–72 days later. **Diet** Mainly ants, their eggs, larvae and pupae, also termites, worms, leeches and other invertebrates. **Habitat** Rainforests, wet and dry sclerophyll forests, woodlands, heaths, shrublands and rocky outcrops in the ranges, slopes and lowlands.

| LENGTH **To 750 mm** | MIDBODY SCALES **22 rows** | STATUS **Low risk.** |

FAMILY **TYPHLOPIDAE**   SPECIES *Ramphotyphlops wiedii*

BROWN-SNOUTED BLIND SNAKE

NON-VENOMOUS, HARMLESS

A worm-like reptile with a blunt, shark-like head indistinct from its neck, a rounded snout and a small, set-back mouth. It is brown to blackish-brown above and creamy-white below with small, tough, shiny, overlapping scales making it impervious to ant bites and helping it to slide through underground tunnels. It has a very short tail with a small, curved spine at the tip used as a digging spike and to anchor it in its tunnels. The tiny, rudimentary eye can detect light and dark, but little else. It lacks an ear aperture, but responds to vibrations. Males are smaller than females. **Behaviour** This very secretive burrowing snake is only occasionally seen above ground on warm, wet nights, and is sometimes found in tightly bundled aggregations under rocks. In dry weather it shelters under rocks, fallen timber and in deep soil cracks. It forages in soft surface soil and among surface debris, and also underground, where it seeks and enters ant nests, looking for ant larvae and pupae. It uses its head and neck to push its way into the soft soil, flicking its pale forked tongue in and out frequently to detect the slightest odour of ants and other invertebrate prey. If alarmed it emits an unpleasant smell from its anal glands. Other glands beneath the head and neck scales secrete an oily fluid which may help it to slide into small crevices. **Development** Females lay 1–8 soft-shelled eggs, and the hatchlings are about 90 mm long. **Diet** Small ants, their larvae and eggs, and other soft-bodied invertebrates. **Habitat** Open woodlands and shrublands, on the plains and undulating country.

LENGTH **To 320 mm**	MIDBODY SCALES **20 rows**	STATUS **Low risk.**

FAMILY **BOIDAE**   SPECIES *Antaresia childreni*

CHILDREN'S PYTHON

NON-VENOMOUS

A small, sinewy constrictor with a narrow head distinct
from its slender neck, a long, squarish snout and bulky jaw
muscles. The eye has a pale iris and a vertical pupil. It is
smooth, shiny, light to reddish-brown above, with or without
a pattern of brown blotches or cross-bars on the back and
sides. A dark streak runs through each eye and the belly is
creamy-white. It has long, backward-curving, razor-sharp teeth,
heat-sensitive pits in the lower lip to detect warm-blooded prey, and a
short spur on each side of the vent (the remnants of hind limbs), used by
the male to hold the female's tail when mating. **Behaviour** This nocturnal
snake is predominantly ground-dwelling, although it occasionally climbs
into trees. It shelters by day in soil cracks, under peeling bark, in tree
hollows, in cavities in termite mounds, and in rock crevices and recesses.
It is very fast and agile, and in some areas lingers around the entrances
to bat caves and clefts, often hanging down, suspended by its tail,
ambushing flying bats as they emerge at dusk. On cool days it basks in
the sun close to its refuge, relying on its drab colouration as camouflage,
and quickly hides if disturbed. In the breeding season rival males engage
in bouts of wrestling, which may include biting, to determine mating
priorities. **Development** They mate from May to August and females lay
2–20 parchment-shelled eggs from September to December. The eggs
stick together and the female coils around them and sometimes shivers
to keep them warm. She protects them until they hatch 46–82 days later.
The hatchlings are about 300 mm long. They are able to breed at 2–3
years and have a lifespan of up to 30 years. **Diet** Small lizards, especially
geckos, frogs, small mammals and some birds. **Habitat** Monsoon forests,
open forests, woodlands and shrublands on tropical plains and slopes
with rocky outcrops.

Unpatterned form

| LENGTH **To 1.1 m** | MIDBODY SCALES **36—49 rows** | STATUS **Low risk.** |

SPOTTED PYTHON

NON-VENOMOUS

A medium-sized python with a stout body, a broad head distinct from its slender neck, and a long, squarish snout. It has smooth, glossy scales, light to medium brown above, with a pattern of dark brown blotches, sometimes forming cross-bars or zigzag stripes. The lower sides are paler and the belly is white or yellowish. A dark streak runs through each eye. It has long, backward-curving, razor-sharp teeth, heat sensitive pits in the lower lip to detect warm-blooded prey, and a short spur on each side of the vent (the remnants of hind limbs), used by the male to hold the female's tail when mating. **Behaviour** This ground-dwelling and semi-arboreal snake is active mainly at night, although it is occasionally seen during the day. It hunts around rocks, in trees and shrubs, and shelters in hollow timber, beneath bark, under rock slabs, in rock crevices, under surface debris, in caves, abandoned burrows and termite nests. Large numbers sometimes congregate at dusk at the entrance to a cave, where they capture bats as they emerge, often hanging in mid-air, suspended by their tail from a rock. It kills its victims by squeezing them in the coils of its body. Males wrestle with each other in the breeding season to determine their mating priorities. **Development** They mate from April to October. Females stop eating about the mid-term of their pregnancy and often warm their bellies in the sun. They lay a clutch of 4–17 eggs, about 39 mm long, in October or November. Females coil around the eggs to protect them from predators, shivering occasionally, to keep them warm. Hatchlings, about 300 mm long, emerge 46–61 days later, and remain in the slit eggs for another 10–48 hours consuming the remaining yolk. **Diet** Mainly small mammals and reptiles, supplemented by small birds and frogs. **Habitat** Rainforest margins, wet and dry open sclerophyll forests and woodlands with rocky outcrops and caves, on the humid to sub-humid plains, slopes and ranges.

| LENGTH **To 1.1 m** | MIDBODY SCALES **35–45 rows** | STATUS **Low risk.** |

| FAMILY **BOIDAE** | SPECIES *Antaresia stimsoni* |

STIMSON'S OR LARGE-BLOTCHED PYTHON

NON-VENOMOUS

A constrictor of Australia's desert regions, it is probably the smallest python in the world. It has a robust body and a long head slightly distinct from its slender neck. The lower lip has small, heat-sensitive pits to detect warm-blooded prey. The scales are smooth, light brown to yellowish-brown above with a contrasting pattern of dark brown blotches, sometimes forming cross-bands. The sides are paler and the belly is white or cream. A dark streak runs through each eye. It has razor-sharp, backward-curving teeth and a short spur on each side of the vent (the remnants of hind limbs), used by the male to hold the female's tail when mating.

Behaviour This is a nocturnal, predominantly ground-dwelling snake, although it sometimes climbs into low shrubs looking for prey, which it ambushes and kills by constriction. It shelters by day under rocks and fallen timber, in soil cracks, tree hollows, in cavities in termite mounds and in abandoned burrows. It is often found in deep rocky crevices and high up on rocky ledges where it preys on roosting bats. It is inoffensive and rarely attempts to bite if handled. If disturbed it curls itself into a ball, hiding its head in the coils of its body.

Development They mate from April to July and females lay a clutch of 6–19 eggs, about 36 mm long, from August to November. The eggs hatch from September to late January. The hatchlings are about 250 mm long and stay in their slit egg cases for up to 24 hours, absorbing the remaining yolk. They have a lifespan of about 8 years. **Diet** Mainly geckos, bats, other small reptiles and mammals, and some frogs.

Habitat Lives in a wide range of arid and semi-arid sites including stony ranges, sand plains and dune fields, also acacia-dominated woodlands and shrublands, hummock grasslands and tree-lined watercourses. **Threats** Habitat degradation due to land clearing and grazing stock, predation by foxes and cats, hunting and illegal collection, loss of tree hollows and fallen timber.

| LENGTH **To 1.27 m** | MIDBODY SCALES **35–49 rows** | STATUS **Vulnerable in NSW.** |

FAMILY **BOIDAE**   SPECIES *Aspidites melanocephalus*

BLACK-HEADED PYTHON

NON-VENOMOUS

A large, non-venomous constrictor with a narrow head indistinct from its neck. It has long, backward-curving, razor-sharp teeth, a round snout and small dark eyes with obscure pupils. The head, neck and throat are glossy jet black. The body is light to dark brown above, patterned with numerous darker cross-bands. The sides are usually paler and the belly is cream to yellow, sometimes with dark blotches. There are no heat-detecting pits in the lower lip, but like other pythons it has a short spur on each side of the vent (the remnants of hind limbs), used by the male to hold the female's tail when mating. **Behaviour** This predominantly nocturnal, ground-dwelling snake forages at night, probing any crevices and holes for prey. It shelters by day in caves, crevices, hollow logs, soil cracks, cavities in termite mounds, grass hummocks and in the burrows of monitor lizards and mammals. It may also dig its own burrow, curving its neck and head in a J-shape and pushing the soil out with its chin. It occasionally basks in the sun on mild days, especially when digesting a large meal, and in cool weather often rests with its black head out of its shelter absorbing heat from the sun. In the breeding season rival males engage in wrestling bouts, entwining their bodies and sometimes biting each other to determine mating priorities. Otherwise, this snake is generally inoffensive, and when threatened or cornered raises its head and hisses, but rarely bites. It kills its prey by constriction. **Development** Mating takes place from July to September and copulation may last for 8 hours. Females fast after mating and frequently warm their bellies in the sun. They lay 6–14 soft-shelled eggs, about 100 mm long, in a sticky clump in October or November, coiling around the clutch for 9–13 weeks to incubate the eggs and protect them from predators, shivering to raise their body temperature on cool days. The hatchlings are about 600 mm long and may stay in their slit egg cases for 24–30 hours, absorbing the remaining yolk. They have a lifespan of at least 7 years. **Diet** Mostly snakes and other reptiles and their eggs, carrion, rarely small mammals and birds. **Habitat** Wet sclerophyll forests, woodlands, shrublands and grasslands in tropical and sub-tropical areas with low rocky ranges and outcrops, rolling hills and plains, from arid inland regions with cracking soils to the coast.

LENGTH **To 3 m**   MIDBODY SCALES **50–65 rows**   STATUS **Low risk.**

FAMILY **BOIDAE**   SPECIES *Aspidites ramsayi*

WOMA PYTHON

NON-VENOMOUS

A large, robust constrictor with a narrow head indistinct
from the neck, small dark eyes with obscure pupils, a rather
pointed snout, a relatively short tail and a thick body. It
is grey, olive to rich reddish-brown above with numerous
darker cross-bands that fade with age. The belly is cream
to yellow with pink or brown blotches. Juveniles often have
black patches on the snout and above the eyes. There are no heat
sensitive pits in the lower lip, but like other pythons it has a short spur on
each side of the vent (the remnants of hind limbs), used by the male to hold
the female's tail when mating. **Behaviour** This predominantly nocturnal,
ground-dwelling snake is usually quiet and shy, and is mostly encountered at
dusk or on warm nights. It shelters in hollow logs, among dense vegetation,
in depressions under grassy hummocks, and in the abandoned burrows
of monitor lizards and mammals (particularly rabbits). In cool weather it
is occasionally seen on the surface during the day. Like other members of
the genus it is able to dig into the ground and bury itself, curving its neck
and head in a J-shape and pushing the soil out with its chin. It hunts at
night, searching for prey in burrows, among spinifex clumps and in trees. It
sometimes wriggles the tip of its tail to lure small animals close enough for it
to strike, and many adults have damaged tails. Prey is squeezed in the coils
of its body, or if caught in the confines of a burrow may be squashed against
the wall until it suffocates. **Development** They mate from May to August
and females lay 4–28 parchment-shelled eggs, about 78 mm long, from
October to December. The eggs stick together and the female coils around
them to incubate them and protect them from predators. Females fast while
incubating their eggs and in cool weather raise their body temperature
by shivering. Hatchlings, about 320 mm long, emerge after an incubation
period of some 60–75 days. **Diet** Reptile eggs, lizards, snakes, small
mammals and some ground birds. **Habitat** Sandhills in the arid
and semi-arid interior with a cover
of hummock grasses
or dense shrubs,
woodlands and
shrublands.
Threats Land
clearing and
degradation
of habitat,
grazing stock,
predation
by foxes and
cats, hunting
and illegal
collection.

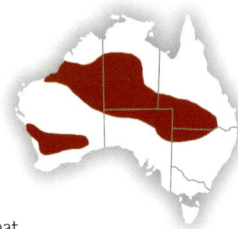

LENGTH **To 2.7 m**   MIDBODY SCALES **50–65 rows**   STATUS **Endangered.**

FAMILY **BOIDAE**   SPECIES *Liasis fuscus*

WATER PYTHON

NON-VENOMOUS

A large constrictor with a strong body and a narrow head
slightly wider than its neck. The scales are water-repellent,
smooth and shiny, iridescent blackish-brown on the back
and below the tail, and yellow on the belly and lower sides.
The throat and lips are creamy white with heat sensitive pits in
the lower lip to detect warm-blooded prey. It has long,
backward-curving, razor-sharp teeth and a short spur on each side
of the vent (the remnants of hind limbs), used by the male to hold the female's
tail when mating. Females are larger than males. **Behaviour** This snake is
predominantly nocturnal, ground-dwelling and semi-aquatic, although it can
climb well if necessary. It shelters by day in hollow logs and stumps, in rock
crevices, among reeds, beneath surface debris, among tree roots and in the
abandoned burrows of goannas and other animals. It emerges at dusk to hunt
for prey near animal trails and in shallow water around their watering places,
sometimes stalking victims for short distances. Its light, slim neck allows it to
strike suddenly with great speed and accuracy. It usually forages in the same
area year round, occasionally travelling 10 km or more following the seasonal
distribution of its prey. In the dry season it hunts for rats in dusty crevices, and
in the wet season it swims around looking for young waterbirds, and is often
found concealed among reeds and other water plants. It takes to the water
when alarmed, staying submerged for long periods if necessary.

Development They mate in the dry season (June to September) and females do
not feed until they lay their clutch of 3–24 parchment-shelled eggs, about
60 mm long, some 3 months later at the beginning of the wet season
(September to November). The eggs are laid in root cavities, under clumps of
earth or in abandoned burrows. Females sometimes coil around the eggs to
incubate and protect them, shivering to increase their body temperature on
cool days. Some 59–80 days later the eggs hatch into 450 mm long hatchlings.
They become sexually mature in their second year and have a lifespan of at least
11 years. **Diet** Mostly native rats, water birds and their
eggs, also young freshwater crocodiles, bandicoots,
wallabies and lizards. **Habitat** Lives in wet
tropical areas around freshwater lagoons,
swamps, creeks, river banks and
dams, in closed and open
forests, and low open
woodlands.

LENGTH **To 3 m**   MIDBODY SCALES **45–55 rows**   STATUS **Low risk.**

FAMILY **BOIDAE**   SPECIES *Liasis olivaceus*

OLIVE PYTHON

NON-VENOMOUS

Often found in the same area as the water python, this is a larger constrictor with a longer, broader head, noticeably distinct from its narrow neck. The scales are glossy with a milky sheen, pale fawn to rich brown or dark olive-brown above, becoming paler on the lower sides, and white or cream on the belly. It has razor-sharp, backward-curving teeth, heat-sensitive pits in the lower lip to detect warm-blooded prey, and a short spur on each side of the vent (the remnants of hind limbs), used by the male to hold the female's tail when mating. **Behaviour** This ground-dwelling, rock-inhabiting snake is mainly nocturnal and shelters by day in deep rock crevices, in caves, in tree hollows, termite mounds and abandoned burrows. It occupies a distinct home range and males travel long distances in June and July to locate and mate with females. In hot weather it often cools down by submerging most of its body in a waterhole, concealed among vegetation or overhanging plants. It ambushes its prey on trails leading to their watering places, or while they are drinking, and kills its victims by constriction. It is a good swimmer and hunts in waterholes, striking from a submerged position. After a large meal it may stay in its shelter for many days while digesting. Males become very aggressive towards each other in the mating season when they compete for mating rights. **Development** They mate from May to July. Females fast after mating and lay a clutch of 8–30 eggs, about 100 mm long, from September to November. The eggs stick together and the female coils around them to incubate and protect them until they hatch 60–87 days later. The hatchlings are about 440 mm long and remain in their slit egg cases with the head protruding for 2–3 days while they absorb the remaining yolk. They do not feed until after their first skin-shedding, about 1 month later. They reach sexual maturity at 3–4 years and have a lifespan of 12 years or more. **Diet** Birds, reptiles and mammals as large as rock wallabies. **Habitat** Monsoon forests, open forests, woodlands, shrublands and hummock grasslands, from arid to humid tropical and sub-tropical districts, preferring rocky hills and ranges, particularly near water.

LENGTH **To 6.5 m**	MIDBODY SCALES **55–80 rows**	STATUS **Low risk.**

FAMILY **BOIDAE**　　　SPECIES *Morelia kinghorni*

SCRUB OR AMETHYSTINE PYTHON

NON-VENOMOUS

The largest snake in Australia, it is named for the milky iridescent sheen of its newly exposed scales after skin-shedding. It is relatively slender with a large, elongated head, broad at the base and distinct from the neck, with large, symmetrical, soft, flexible, plate-like head shields. It has a vertical pupil and heat sensitive pits in the lips to detect warm-blooded prey. It has razor-sharp, backward-curving teeth and 2 short spurs beside the vent (the remnants of hind limbs), used by males to help grasp females when mating. It is olive-yellow to olive-brown above with darker irregular bands, sometimes net-like, forming a single stripe along the lower flanks. The belly is white or cream. **Behaviour** This snake is active primarily at night, although on cold days it is active during the day, and often basks in the sun. It shelters in hollow logs and deep crevices, in tree forks in the canopy, in rock crevices, caves, abandoned buildings and beneath dense vegetation. Younger animals spend most of their time in trees, while older, larger individuals are predominantly terrestrial, although they still readily climb into trees. It is primarily a sit-and-wait predator, often sitting in a coil on the ground next to an area frequented by prey, and striking out sideways when an animal comes close enough. It may also hang from a tree to catch passing prey, or rise up in tall grass to view prey over the top of the grass. Prey is killed by constriction. In winter they move from the dense forests into rocky gorges and lakesides to bask in the sun, and may congregate in open valleys to mate. Males are aggressive in the mating season and often engage in wrestling matches to determine mating rights, rising up and sometimes biting each other. **Development** They mate from June to September after which the females stop feeding and frequently warm their bellies in the sun. A clutch of 5–19 eggs, about 75 mm long, is laid between October and December, and the female coils around the eggs to incubate them, shivering to raise her temperature on cold days. The eggs take 77 to 108 days to hatch. The hatchlings are about 700 mm long and stay in the slit egg cases for up to 2 days, consuming the yolk. **Diet** Mostly mammals, including rats, bandicoots, gliders, fruit bats and wallabies, and occasionally birds. **Habitat** Rainforests, monsoon forests, vine thickets, dry sclerophyll forests and open woodlands of the lowlands, slopes and ranges. Also on coral cays.

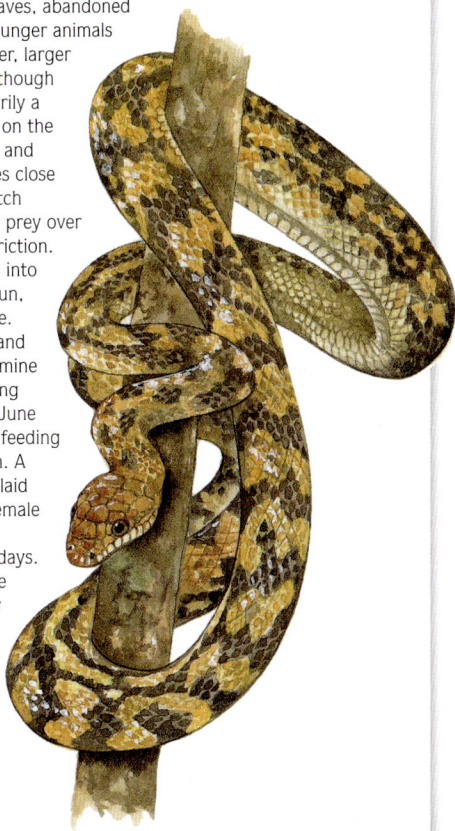

LENGTH **To 8.5 m**　　　MIDBODY SCALES **35–57 rows**　　　STATUS **Low risk.**

FAMILY **BOIDAE**	SPECIES *Morelia spilota*

CARPET OR DIAMOND PYTHON

NON-VENOMOUS

A large python with a long, broadly triangular head distinct from its slender neck, small and fragmented head shields and a vertical pupil. It has heat sensitive pits in the lips to detect warm-blooded prey. There are 3 main subspecies: the carpet python found in most of the continent (M. s. *variegata*) is beige to brown with dark brown, blackish or grey blotches, cross-bands or stripes; the south-western carpet python of WA (M. s. *imbricata*) is typically dark brown with blackish blotches, cross-bands or stripes; the diamond python of coastal NSW (M. s. *spilota*) is typically dark olive above with cream or yellow blotches forming a diamond-shaped pattern. The belly is white, cream or yellow. **Behaviour** An arboreal, terrestrial and rock inhabiting snake, it is active by day in cool weather and by night in warm weather. It is difficult to spot in dappled shade, where it may wait, coiled-up beside a trail for weeks until an animal comes close enough to ambush. It shelters in tree hollows, beneath leaf litter, in animal burrows, caves, deep rock crevices, beneath boulders and in the roof spaces of buildings. Males hunt in a home range of around 45 ha, females use about 20 ha. In cool climates they move to north-facing slopes in winter and often emerge to bask in the sun. In the breeding season adult males travel up to 500 m per day, following the chemical trails of females. Up to 5 males may compete to mate with a female, wrestling to establish dominance, rising high off the ground, intertwining their bodies and sometimes biting savagely. A female may mate with several suitors. Diamond pythons are usually passive when approached, while carpet pythons may hiss and strike. **Development** Females breed in spring every 2–3 years. They fast after mating and warm their bellies in the sun. A clutch of 7–54 eggs, about 55 mm long, is laid in a nest from November to January, and the female coils around her eggs to incubate them, shivering to raise her body temperature on cold days. They hatch 47–81 days later, in January and February. The hatchlings are about 400 mm long and stay in their slit egg cases for about 12 hours consuming the remaining yolk. They have a lifespan of at least 17 years.

Diet Mammals, particularly rats, bandicoots, possums, small wallabies and fruit bats. Also reptiles and some birds.

Habitat Rainforests, forests, woodlands and shrublands, of the coast and ranges.

M. s. spilota

M. s. imbricata

M. s. variegata

LENGTH **To 4 m**	MIDBODY SCALES **46–65 rows**	STATUS **Low risk.**

FAMILY **BOIDAE**   SPECIES *Morelia viridis*

GREEN PYTHON

NON-VENOMOUS

One of Australia's most striking snakes, this moderately small python has a more or less triangular body in cross section and a large head with a bulbous base, distinct from its slender neck. It has a pale iris, a vertical pupil and heat sensitive pits in the lips to detect warm-blooded prey. It has razor-sharp, backward-curving teeth and 2 short spurs beside the vent (the remnants of hind limbs), which the male uses to grip the female's tail when mating. The head scales are granular, while the body scales are smooth. It is dull to bright emerald-green above with a few scattered white scales on the sides. A white or yellow line runs along the spine, crossed by short, light blue bars. Adults are occasionally blue or partially yellow. Juveniles are usually bright yellow, and occasionally orange, brick-red or deep blue. When they are at about 55 cm long and least 6 months old, they begin to turn the green colour of adults. This process can take a few weeks or up to 2 years. **Behaviour** This arboreal snake spends most of its life in the forest canopy. It is active at night and shelters by day on branches, in tree hollows and in large epiphytic ferns. It often rests in a coil with its head in the centre, draped over a branch or vine in dense foliage. In this position it collects drinking water in its body coils when it rains. It moves fairly slowly and searches for prey on the ground or in trees, often taking sleeping animals, or concealing itself in the forest canopy ready to ambush an unwary animal that comes close. Juveniles attract prey close to their resting site by vibrating the thin, bright blue tip of their tail, which acts as a worm-like lure. The victim is grasped by biting and then suffocated by constriction. Males become territorial in the breeding season and often fight each other for mating rights. **Development** Females lay 10–26 eggs, about 38 mm long, in the dry season, from August to November. The eggs are often laid in a tree hollow and adhere to each other. The female coils around the eggs to incubate them, shivering to raise her body temperature. The hatchlings, about 300 mm long, emerge in the wet season, in late November, some 46–68 days later. They begin feeding after their first moult at 10–28 days and are sexually mature at 2–4 years with a lifespan of at least 15 years. **Diet** Reptiles and small mammals, especially rodents, and occasionally birds. Juveniles eat lizards, frogs, birds and eggs. **Habitat** Rainforests, monsoon forests and bamboo thickets on the lowlands, slopes and ranges.

Juvenile

LENGTH **To 2.13 m**   MIDBODY SCALES **50–75 rows**   STATUS **Low risk.**

| FAMILY **ACROCHORDIDAE** | SPECIES *Acrochordus arafurae* |

ARAFURA FILE SNAKE

NON-VENOMOUS

A large and robust aquatic snake with loose skin and tiny, rough, file-like scales. The head is fairly narrow and just distinct from the neck, with small, slightly protruding eyes with vertical pupils, and nostrils on the top. The body is flattened from the side with a narrow, slightly flattened, prehensile tail. It is grey to dark brown above with a broad dark band along the back, and a series of blotches or vague cross-bands extending onto the whitish belly. Females are much larger than males.

Behaviour This snake is strictly aquatic and very agile in the water where its flabby skin flattens into a paddle shape. It is helpless on land and usually stays submerged, raising the top of its head above water to breathe through its valved nostrils. It rests by day close to the banks of waterways, hiding from predatory birds in the shadows, among pandanus roots and floating vegetation. It often travels more than 100 m overnight, moving slowly, hunting within a home range of up to 5 ha. It flicks its deeply forked tongue slowly while hunting, ambushing fish or probing for them in muddy holes. It strikes with remarkable speed, biting prey before constricting it in its coiled body, gripped by its rough skin. It also anchors itself to underwater roots with its prehensile tail, preventing it being washed away and allowing it to seize large, powerful, passing fish. They disperse widely in the wet season floods, and congregate in large numbers in the dry season as the pools shrink, often forming writhing, tumbling balls, comprising 2–7 males attempting to mate with a single female.

Development They mate in the dry season from May to August, and females probably only reproduce every 2–3 years. They give birth to 6–30 live young, about 400 mm long, late in the wet season (February to April). Males mature at about 6 years, females at about 9 years. **Diet** Fish, including large species like barramundi and sleepy cod. **Habitat** Lowland freshwater habitats including pandanus-fringed lagoons, inundated wetlands and the sheltered banks of slow-moving rivers. It freely enters estuarine waters and the sea.

| LENGTH **To 2.5 m** | MIDBODY SCALES **120–180 rows** | STATUS **Low risk.** |

FAMILY **COLUBRIDAE**   SPECIES *Boiga irregularis*

BROWN TREE SNAKE

VENOMOUS BUT NOT DANGEROUS

A moderately large and slender snake, slightly flattened
from the side, with a broad, bulbous head distinct from its
narrow neck, and a long, prehensile tail. It has large eyes
with narrow, vertical pupils. This is a rear-fanged snake with
venom glands connected to fangs at the back of the upper
jaw. It has smooth scales and eastern individuals are pale
brown to reddish-brown above with numerous darker blotches,
sometimes forming narrow cross-bands. West of Cape York there is a cream to
light reddish-brown form with numerous red or reddish-brown cross-bands.
The belly is salmon pink or cream. **Behaviour** This snake is predominantly tree-
dwelling and rock-inhabiting, although it also hunts on the ground. It is active
at night and shelters by day in tree hollows, rock crevices and termite mounds.
Small groups (usually no more than 6 snakes) are often found in caves and
among the rafters of buildings, where their presence is indicated by shed skins.
It moves slowly through the branches of trees, stalking and ambushing its prey,
striking with great speed and accuracy, biting as well as constricting its victims.
If threatened it is usually aggressive, curling its upper body into S-shaped
loops, striking repeatedly and savagely, often with its mouth agape. Adults are
most active in the summer months, and often form aggregations of dormant
snakes in winter. Juveniles are active year round. **Development** They breed in
early summer in the south and in spring in the north, laying a clutch of 3–12
eggs, about 50 mm long, from early spring to early autumn. The eggs are laid
in humid mulch and leaf litter in deep tree hollows, or in deep, humid, rocky
crevices. Females often remain in the area to protect their incubating eggs. The
hatchlings emerge 77–90 days later and are about 400 mm long. **Diet** Small
mammals, birds and their eggs, lizards and some frogs. **Habitat** Rainforests,
mangroves, wet and dry sclerophyll forests, particularly around rocky outcrops,
paperbark swamps and coastal heaths. In the north it also lives in vine forests
and monsoon forests.

Northern form

LENGTH **To 2 m**   MIDBODY SCALES **19–23 rows**   STATUS **Low risk.**

FAMILY **COLUBRIDAE**   SPECIES *Dendrelaphis punctulatus*

GREEN TREE SNAKE

NON-VENOMOUS

A slender snake with a narrow head barely distinct from
its neck. It has large eyes with round pupils and a dark
iris, and a long, prehensile tail. The scales are smooth, but
a long ridge along each side of the belly helps it grip onto
branches. Its colour above varies from grey to olive-green in
NSW and most of Qld, dark brown, black or blue in northern
Qld, golden yellow with a bluish head in the NT and WA. The
skin between the scales is light blue and is visible when the snake is aroused
or flattens its body. The head is often grey or brown and the belly is usually
yellow, but may be white, olive, green or bluish. Females are larger than males.
Behaviour This snake is active mainly during the day, although in the northern
part of its range it can be active at night. It is very fast-moving and alert, with
acute eyesight, and is predominantly tree-dwelling, although it often hunts on
the ground. It actively hunts and chases prey, locating animals by scent, probing
leaves and loose surface cover. Prey is not constricted and is swallowed alive.
It shelters at night among foliage, in tree hollows, rock crevices and narrow
caves, and is often found in suburban areas. In winter, groups of up to 6
individuals often congregate in hollow trees, abandoned buildings and caves. If
cornered it inflates its neck and upper body, displaying the blue skin between its
scales, while hissing loudly, and will bite repeatedly if provoked. It may also emit
a strong-smelling odour from its anal glands. **Development** They mate in spring
and females lay a clutch of 3–15 eggs, about 37 mm long, from September
to December. The eggs are laid beneath bark, under leaf litter, in deep humid
crevices and cavities in rocks, and in hollow tree stumps. They hatch 73–126
days later, and the hatchlings are about 300 mm long. **Diet** Mostly frogs, some
lizards, small mammals, stranded tadpoles and fish. **Habitat** Rainforests, vine
thickets, wet and dry sclerophyll forests, woodlands, coastal heaths and along
inland rivers, from semi-arid to humid areas.

Yellow form

Blue form

| LENGTH **To 2 m** | MIDBODY SCALES **13 (rarely 15) rows** | STATUS **Low risk.** |

| FAMILY **COLUBRIDAE** | SPECIES *Stegonotus cucullatus* |

SLATEY-GREY SNAKE

NON-VENOMOUS

A medium-sized, slender and muscular snake with a narrow, square-snouted head barely distinct from its neck. It has fairly small eyes with a black iris obscuring a vertical pupil. The scales are smooth and glossy with an iridescent sheen. The upperparts are a uniform dark grey, dark brown or black. The lips and lower neck are pale yellow, and the belly is white or cream, sometimes with black blotches and flecks, especially on the lower flanks. Males are larger than females. **Behaviour** This snake is mainly ground-dwelling, although it readily climbs into low foliage and onto steep rock faces, and often enters the water. It hunts around dusk, dawn, at night, and occasionally by day. It shelters beneath logs, rocks, among tree roots, in soil cracks, low vegetation and abandoned burrows, and often enters buildings. This snake lives in a home range of around 3–7 ha throughout the year. It is a good swimmer, and hunts around frog-breeding sites, concealing itself in shrubs and climbing the banks of waterways. It also catches fish underwater in drying pools at the end of the wet season. It suffocates prey, squeezing its victim in the coils of its body. It is very aggressive when aroused and strikes repeatedly and fiercely, while emitting a pungent odour from its anal glands. **Development** They breed in summer and females lay a clutch of 5–16 eggs, about 35 mm long, beneath rotting vegetation from the late dry season to the early wet season. The eggs take 69–110 days to hatch. The hatchlings are about 250 mm long and take 2–3 years to reach sexual maturity. **Diet** Mainly frogs, tadpoles, reptiles and their eggs, with some fish and small mammals. **Habitat** Rainforests, monsoon forests, woodlands and sand dunes in wetter areas along the coastal plains, slopes and ranges, favouring the damp margins of billabongs, swamps, dams, rocky creek beds and even water tanks.

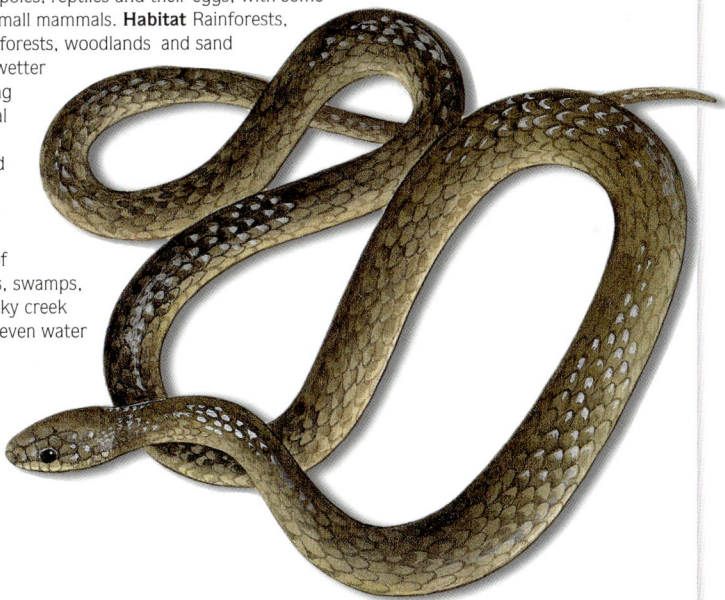

| LENGTH **To 1.5 m** | MIDBODY SCALES **17–19 rows** | STATUS **Low risk.** |

FAMILY **COLUBRIDAE**   SPECIES *Tropidonophis mairii*

KEELBACK SNAKE

NON-VENOMOUS

The name of this snake is derived from its strongly keeled
scales, which are rough with projecting ridges, and help it
to grip slippery surfaces. It is relatively small with a strong
and muscular body, and the head is not distinct from the
neck. It has large eyes with round pupils. Its colour varies
through grey, brown, olive to black above, often with a
number of darker cross-bands, blotches or flecks. The belly is
cream, olive, brown or salmon-coloured. **Behaviour** This semi-aquatic snake
is active by day and night, but is most often encountered at dusk and after
dark, or during the day in cool periods. It shelters around streams, swamps
and lagoons among low vegetation, beneath fallen timber and in abandoned
burrows. It hunts on the ground and in shallow, muddy water, sometimes
staying submerged for as long as 20 minutes, and may bask on low shrubs or
on the surface of warm, shallow water, staying close to the cover of vegetation.
If cornered it rears up with a flattened neck and lunges feebly at its aggressor,
and will bite if provoked. It also emits an unpleasant odour from its anal glands
as a deterrent, and like many lizards is able to discard its tail if grasped by a
predator and grow a new one. In the mating season males wrestle with each
other. **Development** They mate from March to early December and females
lay a clutch of 4–18 eggs, about 25 mm long, from May to November in the
tropical north, and from February to April in the subtropics. More than one
clutch may be produced per season in the tropics. The eggs are laid in a nest
(often used by several females) in humid or moist sites in soil cracks, under
rocks, logs or below rotting leaf litter. They hatch some 43–70 days later. The
hatchlings are about 150 mm long. **Diet** Almost exclusively frogs, including cane
toads and toxic frogs, with no ill-effects. Also some tadpoles, fish, small lizards
and small mammals. **Habitat** Around creeks, rivers, lagoons, swamps and
freshwater lakes with fringing and emerging aquatic vegetation, on
the lowlands, slopes and ranges.

| LENGTH **To 1.2 m** | MIDBODY SCALES **15 (rarely 17) rows** | STATUS **Low risk.** |

FAMILY **ELAPIDAE**   SPECIES *Acanthophis antarcticus*

COMMON DEATH ADDER

DANGEROUSLY VENOMOUS

A distinctive, highly venomous, viper-like snake. It has long, sharp, front fangs, a short, robust body, a broad triangular head quite distinct from its neck, and a thin tail with a small, curved spur at the tip. Its eyes are small and set relatively high on its head, with a vertical pupil and a pale iris. It has smooth to slightly rough scales, and its colour varies from blackish to pale grey or rich reddish-brown above, often with irregular dark or light cross-bands. The belly is cream or grey with dark brown or grey flecks and spots. The tip of the tail is usually white or cream, and the lips are prominently barred. Females are much larger than males.

Behaviour This sluggish, primarily ground-dwelling snake, is most active on warm afternoons and nights, and shelters in hollow logs, beneath vegetation and ground cover. It moves slowly and stealthily, occasionally side-winding, climbing over or around dry leaf litter to avoid making any noise. Unlike most other venomous snakes it does not actively search for prey, but waits for prey to come to it. Concealed beneath vegetation or half buried in sand or leaf litter, it lies motionless, loosely coiled with its tail in front of its snout, and lures victims closer by wriggling the tip of its tail, which is often pale or yellow, to mimic a worm or caterpillar. It strikes with lightning speed, injecting highly toxic venom. When agitated it flattens its whole body, and if threatened it forms a rigid coil and flicks from side-to-side, sometimes striking repeatedly. It is not generally aggressive, however, and usually tolerates being stepped on momentarily. On warm spring nights males roam around searching for females to mate with. **Development** They mate in spring and sometimes in autumn. Females usually reproduce every second year unless there is abundant food and warm weather. They give birth to a litter of 2–42 live young, about 150 mm long, between December and April. Males are sexually mature at 2 years, females at 3.5 years, and have a lifespan of at least 10 years.

Diet A wide variety of vertebrates. Adults eat mainly small mammals and birds, while juveniles feed mostly on lizards and frogs. They drink from pools on leaves.

Habitat Rainforests, wet sclerophyll forests, woodlands, shrublands and coastal heaths on the plains, slopes and lower ranges.

Threats Habitat loss and disturbance by land clearing and grazing, cane toad poisoning.

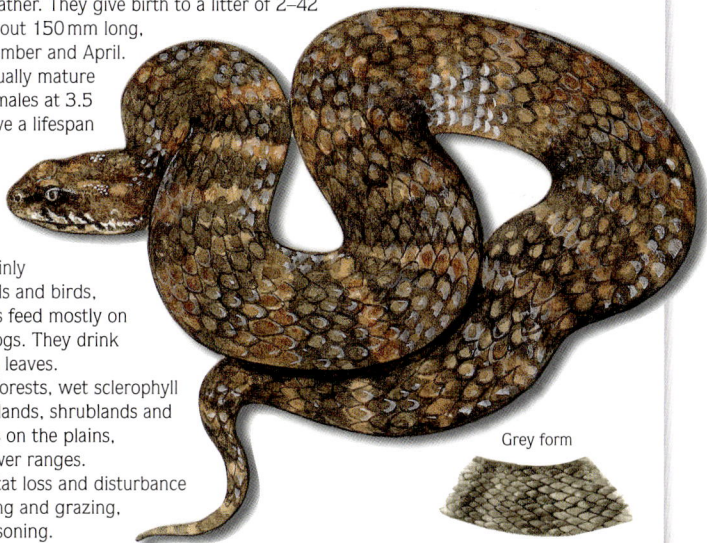

Grey form

LENGTH **To 1.05 m**   MIDBODY SCALES **21 (rarely 23) rows**   STATUS **Threatened in Vic.**

| FAMILY **ELAPIDAE** | SPECIES *Acanthophis praelongus* |

NORTHERN DEATH ADDER

DANGEROUSLY VENOMOUS

A highly venomous, viper-like snake, similar to the common death adder. It differs mainly in its smaller size, lighter build and rough head scales, which usually protrude above the eye. It has long and sharp front fangs, a robust, short body with a broad triangular head quite distinct from its neck, and a thin tail with a spiny tip. Its small eyes are set relatively high on the head and have vertical pupils and pale irises. The scales are smooth or slightly rough on the back. It varies in colour from grey to brownish-red above, usually with distinct light and dark cross-bands, and has a whitish belly. Females are larger than males. **Behaviour** This secretive snake spends most of the day half buried in sand, soil or leaf litter, usually in spinifex hummocks or beneath trees or shrubs. It moves around on warm afternoons and nights, and avoids making a noise by climbing over or around dry leaf litter. Unlike most other venomous snakes it does not actively search for prey but ambushes prey from a partly concealed position, lying motionless in the leaf litter with its body loosely coiled and the tip of its tail held up close to its head. When potential prey approaches it wriggles the tip of its tail which looks like a worm or caterpillar, luring the animal into range and striking with lightning speed and great accuracy, injecting highly toxic venom. It relies on colouration to avoid detection, and if approached it normally remains motionless. It is not generally aggressive, and when concealed usually tolerates being momentarily stepped on. If strongly provoked it flattens its body into a tense coil, flicks its head from side-to-side before striking fast and often repeatedly. **Development** Females usually reproduce every second year, giving birth to up to 20 live young from February to April. The newborn are about 130 mm long. Males reach sexual maturity at 12 months, females at 18 months.
Diet Mainly lizards and frogs, some small birds and small mammals.
Habitat Wet and dry sclerophyll forests and woodlands, grasslands, rocky ranges and rocky outcrops along the plains, slopes and ranges, usually in undisturbed habitats.
Threats Habitat loss and disturbance by land clearing and grazing, cane toad poisoning.

| LENGTH **To 790 mm** | MIDBODY SCALES **23 (rarely 21) rows** | STATUS **Near threatened in the NT.** |

FAMILY **ELAPIDAE**   SPECIES *Acanthophis pyrrhus*

DESERT DEATH ADDER

DANGEROUSLY VENOMOUS

A highly venomous snake with long, sharp, front fangs and
a short, stocky body. It is far more slender than the other
death adders, and has a broad triangular head quite distinct
from its neck and a thin tail ending in a small, soft spine. The
eyes are small and high-set with vertical pupils and pale irises.
It is bright reddish-brown with lighter cross-bands above
and cream or reddish below, with rough scales on the head and
back. Females are larger than males. **Behaviour** This ground-dwelling snake is
predominantly nocturnal and shelters by day in and beneath hummock grasses
and in abandoned burrows. It emerges at night and on warm afternoons,
ambushing its prey while partly concealed beneath vegetation or half buried
in loose sand, with its tail tip held up in front of its snout. Prey is lured into
range by wriggling the tail tip, which resembles a caterpillar or worm. The
snake strikes with great speed and accuracy, subduing its victim quickly with
its powerful venom. In warmer weather it hunts for lizards in open areas,
actively stalking its prey and moving very quickly when necessary. This snake
is not aggressive and when concealed usually tolerates being momentarily
stepped on. If provoked, however, it forms a rigid coil with its body flattened,
bringing the pale cross-bands into sharp contrast, and flicks its upper body
from side-to-side. If the aggressor does not retreat it may strike repeatedly.
Development They mate in October and November and females give birth to a
litter of 8–18 live young in mid to late summer or early autumn. The newborn
are about 140 mm long. **Diet** Predominantly lizards, especially skinks and
dragons. **Habitat** Arid regions, in hills, ranges, sandy ridges, stony flats and
rocky outcrops, where there are hummock grasslands
and usually acacia shrubs or sparse, low trees.

| LENGTH **To 750 mm** | MIDBODY SCALES **19–21 rows** | STATUS **Low risk.** |

| FAMILY **ELAPIDAE** | SPECIES *Austrelaps ramsayi* |

HIGHLAND OR ALPINE COPPERHEAD

DANGEROUSLY VENOMOUS

A medium-sized snake with a fairly narrow head, barely distinct from its neck. The eye has a round, dark pupil and a pale iris. It is quite robust, strongly muscled, smooth and semi-glossy above, and is very similar to the lowland copperhead, although it has much more prominent black and white bars on the upper lips. It varies from pale brown to grey or black above with a coppery head and a light grey to creamy-yellow belly. Males are much larger than females. **Behaviour** This ground-dwelling snake is active during the day and on warm nights. It is quite tolerant of cold weather, and on cool days when most other reptiles are inactive it may be seen coiled up in an open area basking in weak sunlight. It is generally secretive, however, and avoids encounters with humans. It hunts around waterways and shelters under rotting timber, in tussock grasses and other dense vegetation, in deep burrows and below large flat rocks, sometimes in small groups. In winter it stays in its shelter and becomes dormant. In the breeding season rival males sometimes gather around a reproductive female and engage in wrestling matches to determine which will mate with the female. It is not usually aggressive, but if cornered or provoked it rises in a low curve, flattens its neck, hisses loudly and often flicks from side-to-side or thrashes menacingly, but is reluctant to bite. **Development** They mate in autumn and spring, and most females breed every second year, giving birth to 3–32 live young from late January to March. The newborn are about 170 mm long and become sexually mature at about 2 years of age. **Diet** Mainly frogs and small skinks, with other lizards, snakes, small mammals and some insects. They occasionally eat the young of their own species. **Habitat** Wet and dry sclerophyll forests, woodlands, montane heaths and tussock grasslands in the cool, high tablelands and ranges, around waterways and seepages.

| LENGTH **To 1.2 m** | MIDBODY SCALES **15–17 rows** | STATUS **Low risk.** |

FAMILY **ELAPIDAE**	SPECIES *Austrelaps superbus*

LOWLAND OR COMMON COPPERHEAD

DANGEROUSLY VENOMOUS

A medium-sized snake with a fairly narrow head, barely wider than its neck, a dark round pupil and a pale iris. It is fairly robust and strongly muscled, and closely resembles the highland copperhead, although the bars on the upper lip are obscure or much less prominent. The scales are smooth and semi-glossy. The colour ranges from deep brick red to brown, coppery red-brown through grey to almost black above, and some specimens have a dark narrow stripe along the spine. The belly is cream, yellow, pink or orange. Males are much larger than females. **Behaviour** This predominantly ground-dwelling snake is mainly active during the day, although it also emerges in the evenings and at night in hot weather. It shelters under or inside rotting logs, in abandoned animal burrows, in clumps of vegetation, under boulders and sheets of roofing iron and other surface debris. In winter it stays in its shelter and becomes dormant. It avoids encounters with humans, and is generally shy and secretive, although large numbers often gather around marshes where prey is abundant. It is not aggressive, and will always flee if possible. If cornered or provoked it flattens its upper body to present the largest possible aspect to the aggressor, rises up in a shallow curve, hisses loudly and often flicks itself from side-to-side or thrashes menacingly, but is usually reluctant to bite. In the breeding season rival males engage in wrestling matches to determine which will mate with a reproductive female. **Development** They mate in late summer and autumn, and on warm winter days. Most females breed every second year, storing the sperm over winter, and give birth to 6–30 live young from January to March. The newborn are about 180 mm long and reach sexual maturity at 2 years of age. **Diet** Mainly frogs and lizards, with the occasional smaller snake (including its own species), insects, small mammals and birds. **Habitat** Wet and dry sclerophyll forests, woodlands, heathlands, shrublands, tussock grasslands and sedgelands, in cool to cold coastal sites and slopes to about 1000 m. It also lives around watercourses, swamps and seepages, and in heavily disturbed areas around dams, ditches and along road verges.

LENGTH **To 1.75 m**	MIDBODY SCALES **15 (rarely 17) rows**	STATUS **Low risk.**

| FAMILY **ELAPIDAE** | SPECIES *Cacophis harriettae* |

WHITE-CROWNED SNAKE

VENOMOUS BUT NOT DANGEROUS

A small, moderately slender snake with a white or pale yellowish collar on the back of its neck forming a stripe around the side of the head and snout. It has a rounded head indistinct from its neck, and small eyes with a pale, mottled iris and vertical pupil. The scales are smooth, glossy and water-repelling. It is dark grey-brown or dark steely-grey above and lighter grey below. Females are larger than males.

Behaviour This is a secretive, predominantly nocturnal, ground-dwelling, snake. It shelters by day under rocks, beneath rotting timber, and in piles of rotting vegetation, and is often found in compost heaps in the outer suburbs of Brisbane. It hunts in open areas at night, ambushing prey that comes within range, and searching burrows and other shelter sites. It uses both venom and constriction to kill its victims. If cornered or provoked it confronts its aggressor, holding its head high off the ground with its snout pointing down and the back of its head flattened. The dark head and contrasting paler crown resemble a gaping mouth. It also thrashes around threateningly and may strike with its mouth closed, but it rarely attempts to bite. **Development** They breed year round with mating peaks in spring and summer. Females lay 2–10 eggs, about 22 mm long, mostly in early summer, and may produce more than one clutch per season. The eggs hatch from late summer to early autumn, after incubating for 10 to 12 weeks, and the hatchlings are about 80 mm long. Males are sexually mature at about 20 months, females at about 32 months. **Diet** Small lizards (mainly skinks), blind snakes, lizard eggs and frogs. **Habitat** Wet and dry sclerophyll forests and woodlands, rainforests, coastal heathlands, moist shrublands and well-watered gardens along the coastal plains and ranges, in gullies, valleys and on the lower slopes. **Threats** Clearing and fragmentation of habitat due to agriculture and forestry activities.

| LENGTH **To 560 mm** | MIDBODY SCALES **15 rows** | STATUS **Vulnerable in NSW.** |

FAMILY **ELAPIDAE**   SPECIES *Cacophis krefftii*

DWARF CROWNED SNAKE

VENOMOUS BUT NOT DANGEROUS

A small, very slender snake with a narrow white collar across the back of its neck, extending around the side of the head and snout. It has a rounded head indistinct from its neck, small eyes with a pale, mottled iris and a vertical pupil. It has a small mouth and short fangs. The scales are smooth, glossy and water-repelling. It is dark steel-grey to black above with a white belly and a black stripe underneath the tail. Females are larger than males. **Behaviour** This ground-dwelling snake is active mainly at night and is quite secretive, sheltering by day under rotten timber, beneath tree bark, in deep leaf litter and below flat rocks. It may be found hunting for food in open areas at night, searching shelter sites and among the leaf litter for prey. It also ambushes any prey that ventures into its shelter, killing victims using constriction as well as venom. When cornered or threatened it attempts to bluff its aggressor with a threat display, holding its head high off the ground with the back of the head flattened so that the pale crown resembles a gaping mouth. It thrashes about threateningly and may strike with its mouth closed, but rarely attempts to bite. **Development** They mate year round with peaks in spring and summer. Females lay a clutch of 2–5 elongated eggs, about 27 mm long, from November to December. The eggs hatch from February to March, 12–14 weeks after laying, and the hatchlings are about 100 mm long. They reach sexual maturity at about 20 months. **Diet** A variety of vertebrates, mainly skinks, other small reptiles and reptile eggs. **Habitat** Rainforests, wet sclerophyll forests, coastal shrublands, estuarine wetlands, along the coastal slopes and ranges, in valleys, gullies and around watercourses. Also found occasionally in well-watered urban areas under compost, logs and stones, and in banana plantations.

| LENGTH **To 350 mm** | MIDBODY SCALES **15 rows** | STATUS **Low risk.** |

FAMILY **ELAPIDAE**	SPECIES *Cacophis squamulosus*

GOLDEN-CROWNED SNAKE

VENOMOUS — LARGE INDIVIDUALS MAY BE HARMFUL

A small snake with a prominent light fawn, brown or yellowish band around the side of the head and snout, often flecked with dark brown, extending onto the back of the neck, but not forming a complete collar. It has an angular head distinct from its slender neck, small eyes with a yellowish, mottled iris and a vertical pupil, a small mouth and short fangs. The scales are smooth, glossy and water-repelling. It is dark brown to steel-grey above. The belly is pink to red with black blotches that coalesce to form a black stripe below the tail. Females are larger than males. **Behaviour** This is a secretive, nocturnal, ground-dwelling snake that shelters during the day beneath fallen timber, under bark, below well-embedded flat rocks, among dense vegetation or rotting leaf litter. It forages in open areas at night, searching in likely shelter sites for sleeping prey, and is often active on nights too cold for most other snakes. Its victims are detected by scent and subdued by constriction as well as venom. If confronted this snake generally tries to escape, but if cornered it adopts a fierce threat display, raising its head high off the ground with its neck held in a tight S-shaped curve. It flattens its head and tilts it down to display the head markings which resemble a gaping mouth. It thrashes wildly, but rarely bites, although it may strike repeatedly with its mouth closed. **Development** They mate year round with peaks in spring and summer, and females lay 2–15 elongated eggs, about 27 mm long, usually from October to January. The eggs hatch 10–14 weeks later, usually in late summer and autumn, and the hatchlings are about 160 mm long. Males are sexually mature at 20–32 months, females at 32–44 months. **Diet** Lizards (mainly skinks, often only their tails), lizard eggs, blind snakes and small frogs.

Habitat Rainforests, wet sclerophyll forests, woodlands and shrublands along the coastal plains, slopes, valleys and ranges. Abundant in sandstone areas in the south and in deep forests in the north. Also found in suburban gardens beneath compost heaps, large stones and logs.

LENGTH **To 850 mm**	MIDBODY SCALES **15 rows**	STATUS **Low risk.**

FAMILY **ELAPIDAE**   SPECIES *Cryptophis nigrescens*

EASTERN SMALL-EYED SNAKE

VENOMOUS — POTENTIALLY DANGEROUS

A small snake with a broad, flattened head just distinct from its neck, and small, beady eyes with a round pupil and black iris. It has smooth, glossy scales, steely bluish-black above and cream to bright pink below, often with dark blotches or flecks. It is similar to the juvenile red-bellied black snake, except for its small eyes and lack of pink on the sides. Males are larger than females. **Behaviour** This secretive, ground-dwelling snake is usually only found on the surface at night. It searches in hollows, among ground litter and other likely sheltering sites for sleeping lizards and other animals. It shelters by day under rocks, stones, logs, in rocky crevices, under ground litter, beneath pieces of bark on the ground, and in termite mounds, where it captures any lizards that venture into the mound. Aggregations of up to 29 snakes have been found sheltering together in winter, their bodies entwined in a large knot. If cornered or provoked it flattens its body into stiff, open coils, hisses loudly and thrashes about wildly, and may strike, although it usually avoids biting humans. The toxicity of the venom can be quite high in some populations, and one human fatality has been recorded. Rival males wrestle each other in the breeding season for the opportunity to mate. **Development** They mate in late autumn, winter and spring. Females give birth to 1–8 (usually 4–5) live young in February or March in the south, and from October to February in the north. The newborn are about 155 mm long and reach sexual maturity in 2–3 years. **Diet** Lizards, mainly skinks, dragons, blind snakes, and occasionally other small snakes, frogs and lizard eggs. **Habitat** Coastal heathlands, woodlands, wet sclerophyll forests and rainforests, also agricultural and grazing lands. It favours rocky outcrops in humid areas along the coast and ranges.

LENGTH **To 1 m**	MIDBODY SCALES **15 rows**	STATUS **Low risk.**

| FAMILY **ELAPIDAE** | SPECIES *Demansia vestigata* |

BLACK WHIP SNAKE

VENOMOUS — LARGE INDIVIDUALS MAY BE DANGEROUS

A long, slender snake with a long, whip-like tail and a deep, narrow head, barely distinct from its neck. There is a slight ridge between the nostril and eye, and the eyes have a very pale iris and a round pupil. The scales are smooth and matt, light olive-brown to black above, with darker edges forming a net-like pattern in pale individuals. The side of the head has faint contrasting light and dark markings. The belly is yellowish-grey and the underside of the tail is reddish. Males are larger than females. **Behaviour** One of Australia's fastest snakes, it is able to move at more than 10 kph, chasing and capturing lizards on the run. It is active by day and also on wet, hot nights, and forages year round, using its acute eyesight to locate prey. It hunts mostly in open areas and shelters at night in low vegetation, beneath leaf litter and other surface debris (including roofing iron). While on the move it often keeps its head high off the ground to see into the distance, pausing and bobbing up and down from time to time. Prey is subdued by the use of constriction as well as venom. A nervous snake, it relies on speed to escape danger, and is very reluctant to bite a human. In the breeding season rival males engage in wrestling bouts to determine which will mate with a reproductive female. **Development** They breed year round. Females lay a clutch of 4–13 eggs, about 33 mm long, under a rock or in a crack in the soil. The hatchlings emerge some 69–72 days later and are about 275 mm long. They reach sexual maturity at about 20 months. **Diet** Lizards, mainly skinks and dragons, and frogs when available. **Habitat** Prefers dry, rocky sites in heathlands, and open forests and woodlands with a grassy understorey.

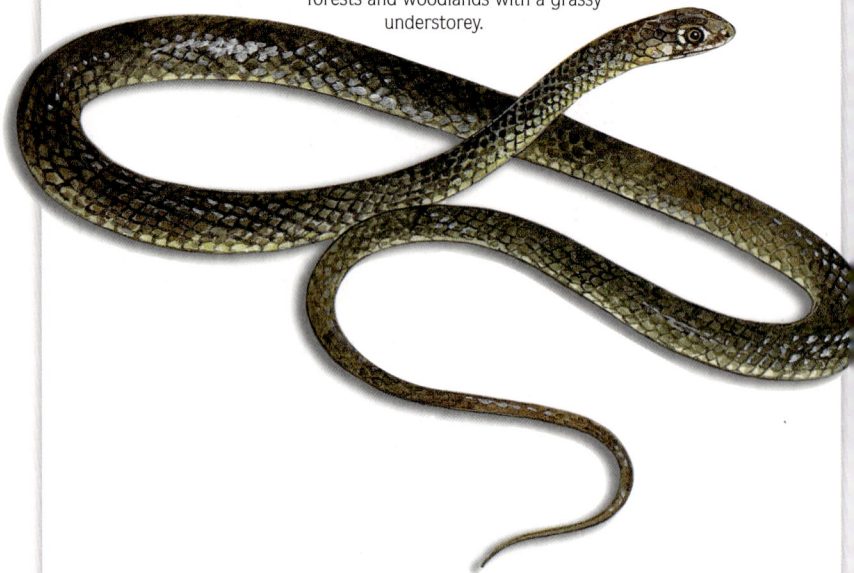

| LENGTH **To 2 m** | MIDBODY SCALES **15 rows** | STATUS **Low risk.** |

| FAMILY **ELAPIDAE** | SPECIES *Demansia psammophis* |

YELLOW-FACED WHIP SNAKE

VENOMOUS — LARGE INDIVIDUALS MAY BE HARMFUL

A slender, medium-sized snake with a long, whip-like tail and a deep, narrow head, barely distinct from its neck. It has very large eyes with a pale iris and round pupil, and a slight ridge running from the nostril to the eye. A dark, narrow, yellow-edged bar runs between the nostrils, and there is a dark, pale-edged, comma-like mark between the eye and mouth. The scales are smooth and matt, and very variable in colour. South eastern individuals are often light grey or olive-grey above. In SA and WA they are grey or greenish above with a dark, net-like pattern. North-eastern populations are usually grey with a reddish flush on the back. The belly is grey-green to yellowish. This snake is able to change colour subtly over a period of a few minutes. Males are larger than females. **Behaviour** This fast, nervous snake is able to move at more than 10 kph, chasing and catching lizards on the run. It is active by day and feeds year round, using its acute eyesight to detect its prey. It generally hunts in open areas and kills prey by venom or by constricting a victim in body coils. Surface debris, low vegetation, crevices and hollows in fallen timber are used as shelter sites at night. It returns to the same area every year to breed, often congregating in large numbers, and up to 20 adults use the same shelter in winter. In south-western areas it is inactive in winter. Males engage in ritualised wrestling matches in the breeding season, with rivals twisting around each other and sometimes biting. It relies on its speed to escape danger, and flees quickly when disturbed. It is reluctant to bite if threatened, but will not hesitate to bite if caught. **Development** They mate in late winter and spring, and females lay a clutch of 3–9 eggs in early summer in the southern part of its range. Females sometimes use communal nests in deep soil cracks, rock crevices or under a large rock exfoliation (600 eggs were found in one site, some already hatched, indicating its use over several seasons). The hatchlings are about 170 mm long and are sexually mature at about 20 months. **Diet** Small skinks, geckos and dragons, with some frogs and reptile eggs. **Habitat** Rainforest margins, dry and wet sclerophyll forests, woodlands, heathlands and hummock grasslands, from the deserts to the coast, staying close to watercourses in desert regions.

SA and WA form

| LENGTH **To 1 m** | MIDBODY SCALES **15 rows** | STATUS **Low risk.** |

FAMILY **ELAPIDAE**   SPECIES *Demansia torquata*

COLLARED WHIP SNAKE

VENOMOUS BUT NOT DANGEROUS

A very slender, long-tailed snake with a deep, narrow head, barely distinct from its neck, and a slight ridge running from the nostril to the eye. The eyes are very large with a pale iris and round pupil, and have a distinctive, dark, comma-like mark around them, running to the corner of the mouth, edged in yellow. A series of black and yellow bands cross the back of the head and neck, and a pale-edged dark bar runs between the nostrils. The scales are smooth and matt. It is grey-brown to olive brown above, sometimes with a reddish tinge. The throat is yellowish with grey streaks, and the belly varies from yellow to dark grey, sometimes tinged with red. This species has been misapplied to other *Demansia* species found in central and western Qld, the NT and NSW. **Behaviour** This swift-moving snake is active by day and probably on warm nights. Like other whip snakes it has acute eyesight which it uses to detect prey, chasing and capturing lizards on the run, moving at more than 10 kph to catch fast-moving skinks. Prey is subdued by constriction and venom. It forages throughout the year, favouring open areas. At night it shelters beneath fallen timber, flat rocks and surface debris, in deep soil cracks, rock crevices and beneath low vegetation. Although venomous it is regarded as harmless to humans. If threatened it attempts to escape, and is very reluctant to bite. Males compete for the chance to mate in the breeding season, engaging in wrestling bouts, with rivals coiling around each other and sometimes biting. **Development** They breed in spring in the south, and probably at any time of the year in the north. Females lay a clutch of 2–8 eggs, and the hatchlings are about 170 mm long. **Diet** Mainly skinks, also dragons, geckos and other reptiles when available. **Habitat** Open forests, grassy woodlands, heathlands and hummock grasslands, along the coast and ranges.

| LENGTH **To 636 mm** | MIDBODY SCALES **15 rows** | STATUS **Low risk.** |

FAMILY **ELAPIDAE**   SPECIES *Denisonia devisi*

De Vis' Banded Snake
Mud Adder

VENOMOUS — LARGE INDIVIDUALS MAY BE HARMFUL

A small, robust snake of the flood plains and western slopes, sometimes mistaken for a small death adder or eastern tiger snake. It has a broad, flattened head, distinct from its neck, and large eyes with a vertical pupil and conspicuous pale iris. It is smooth, yellowish-brown, olive-brown or reddish-brown above, with a series of broad, irregular dark brown cross-bands. The head is dark brown with lighter flecks and conspicuously barred lips. The belly is white or cream. **Behaviour** This nocturnal, fairly sluggish, ground-dwelling snake shelters by day under fallen timber, beneath ground litter and in soil cracks and other deep cavities. It emerges at night and hunts actively on warm nights, often near water and on black-soil river flats, searching out and stalking its prey. In addition to a typical snake movement it is also able to move by side-winding. It is pugnacious, and if provoked it flattens its body into stiff curves, thrashes wildly and will strike repeatedly when the opportunity arises. **Development** Females give birth to 3–9 live young per litter. The newborn are about 110 mm long. **Diet** Principally frogs with some skinks, soft-skinned geckos and the occasional invertebrate. **Habitat** Woodlands and scrublands, favouring alluvial flats in low-lying, moist sites in semi-arid areas. **Threats** Land degradation due to grazing stock and agricultural practices.

| LENGTH **To 600 mm** | MIDBODY SCALES **17 rows** | STATUS **Critically endangered in Vic.** |

| FAMILY **ELAPIDAE** | SPECIES *Drysdalia coronoides* |

WHITE-LIPPED SNAKE

VENOMOUS BUT NOT DANGEROUS

A relatively small and slender snake with a deep, narrow head, slightly distinct from its neck, and a conspicuous thin white stripe along its upper lip from the snout to the side of the neck. It has large eyes with a round pupil and dark iris. The scales are smooth and matt. The upper colour varies from light grey to olive-green, brick red or almost black. The belly is yellow, cream or pink. Tasmanian individuals are generally thicker-bodied and dark green. **Behaviour** This is the most cold-adapted of all the Australian snakes, with populations living in alpine areas. It is a ground-dweller, and is active mainly at night, although it sometimes forages during the day and basks in the sun. It shelters beneath dense matted vegetation, especially tussock grasses, in deep leaf litter, under rotting timber and under well-embedded rocks. It is an active predator and moves around searching for and stalking prey. It flattens its body when basking or provoked, but will not bite unless roughly handled. In the breeding season females congregate in colonies and often bask in the sun together, while the males compete for the chance to mate, engaging in ritualised wrestling bouts. **Development** Females on the mainland give birth every year, while those in Tasmania produce a litter every 2–3 years. They mate in spring and females give birth to 2–10 live young from late summer to mid autumn. The newborn are about 110 mm long and mature at about 32 months. **Diet** A variety of vertebrates, mainly skinks, with some lizard eggs, frogs and small mammals. **Habitat** Dry sclerophyll forests, the margins of wet sclerophyll forests, woodlands, heathlands, tussock grasslands and sedgelands, in wet or seasonally waterlogged sites, on cool temperate to temperate coastal plains, slopes and ranges, to alpine plateaux. Also found in suburban gardens.

| LENGTH **To 500 mm** | MIDBODY SCALES **15 rows** | STATUS **Low risk.** |

FAMILY **ELAPIDAE**   SPECIES *Furina diadema*

RED-NAPED SNAKE

VENOMOUS BUT NOT DANGEROUS

A small, slender snake with a slightly flattened head, barely distinct from its neck. It is distinguished by its glossy black head and neck and a bright red or orange crescent, oval or diamond-shaped patch on the back of the head. The eyes are small and black with a round pupil, and there is a white streak along the upper lip. The scales are smooth, glossy and water-repelling. The colour above is rich reddish-brown, edged in black to form a net-like pattern. The belly is white or cream. Females are larger than males. **Behaviour** This ground-dwelling, nocturnal snake shelters by day beneath rocks, logs, leaf litter, in abandoned burrows, termite mounds and in deep cracks in the ground. It is sometimes found sharing a shelter with several others. It feeds year round, but is more active in the warmer months. It searches for skinks in their night-time retreats, slipping stealthily into cracks, crevices, burrows and other confined spaces, aided by its slender head and fore-body, subduing its victims by constriction as well as venom. If threatened or alarmed it raises its fore-body high off the ground, holds itself stiffly and thrashes around, striking repeatedly, but usually with its mouth closed, only biting if severely provoked. **Development** They mate in late spring or early summer, and females lay 1–5 eggs per clutch, possibly producing more than one clutch per year in Qld. The eggs hatch in January or February in cooler areas, and the hatchlings are about 120 mm long. They reach sexual maturity in the year following their birth. **Diet** Skinks. **Habitat** Dry sclerophyll forests, woodlands, shrublands and tussock grasslands, from the humid coast to arid inland areas, on the ranges, slopes and plains, in sites with rocky outcrops. It is often associated with termite and ant colonies. **Threats** Habitat disturbance and fragmentation due to land clearing, grazing stock and agricultural practices.

LENGTH **To 400 mm**	MIDBODY SCALES **15 rows**	STATUS **Vulnerable in Vic.**

FAMILY **ELAPIDAE**	SPECIES *Furina ornata*

ORANGE-NAPED SNAKE

VENOMOUS BUT NOT DANGEROUS

A small, slender snake with a slightly flattened head, barely distinct from its neck. It is distinguished by a broad, bright orange or yellow band running across the back of its neck, separating its dark brown to black head from its similarly coloured neck. It has small black eyes with round pupils and pale cream lips. The scales are smooth, glossy and water-repelling. They are pale orange to reddish-brown and edged with black to create a net-like pattern over the back. The belly is pale cream. Females are larger than males. **Behaviour** This ground-dwelling, nocturnal snake shelters by day in deep cracks in the ground, in the disused burrows of other animals, beneath logs, in dead vegetation and sometimes under rocks. It forages at night, searching for skinks, using its slender head and fore-body to probe into confined spaces where they shelter at night. Its victims are subdued by constriction as well as venom. If threatened or disturbed it rises high off the ground, thrashes around menacingly and strikes with its mouth closed, but does not bite unless severely provoked. **Development** Females lay a clutch of 1–6 eggs in summer, and in good conditions may produce more than one clutch per year. The hatchlings are about 120 mm long. **Diet** Skinks. **Habitat** Open forests and woodlands, shrublands and hummock grasslands, from the humid coast to arid inland areas, on the plains, slopes and ranges, with deep cracking soils.

LENGTH **To 700 mm**	MIDBODY SCALES **15–17 rows**	STATUS **Low risk.**

BLACK-BELLIED SWAMP SNAKE
MARSH SNAKE

VENOMOUS BUT NOT DANGEROUS

A medium-sized, moderately robust snake with a narrow head barely distinct from its neck. The eyes have round pupils surrounded by a gold rim and a dark iris. Narrow, white or yellow, dark-edged streaks run from the eye to the side of the neck and along the upper lip. The scales are smooth and semi-glossy, usually brown or olive-brown above, varying to black in coastal NSW populations, and pink in some parts of Qld. The belly is black or dark grey, varying to salmon or cream in Qld. **Behaviour** This ground-dwelling snake hunts during the day and in the evenings, and also on warm, wet nights. It shelters beneath fallen timber, under rocks, in dense mats of vegetation, and in warmer weather among dense aquatic vegetation. It searches for skinks in their shelters, flattening its head and body so that it can squeeze into cracks and crevices. Large numbers congregate in suitable habitats and sometimes around rubbish dumps. In the breeding season rival males compete for the chance to mate, intertwining their bodies and wrestling with each other. They are generally quite docile, and if threatened rise up and flatten their head and fore-body, but do not bite unless severely provoked or roughly handled. The bite is painful but not dangerous. **Development** Mating begins in autumn and continues through winter (in warm weather) and spring. Females give birth to 3–15 live young from January to April. The newborn are about 135 mm long with velvety dark heads and bright colours. They become sexually mature at about 12 months. **Diet** A variety of invertebrates including lizards, especially skinks, lizard eggs, frogs, tadpoles and other snakes. **Habitat** Rainforests, wet and dry sclerophyll forests and woodlands, shrublands, heaths, tussock grasslands and sedgelands, along the humid coast and ranges. It is abundant around creeks and swamps, but is also found on dry rocky ridges, beach dunes and in well-watered gardens.

| LENGTH **To 914 mm** | MIDBODY SCALES **17 rows** | STATUS **Low risk.** |

PALE-HEADED SNAKE

VENOMOUS — LARGE INDIVIDUALS MAY BE HARMFUL

A small, slender, well-muscled, tree-dwelling snake with a broad, flattened head distinct from its neck, large eyes and round pupils. The top of the head is grey, often with prominent spots, and there is a broad white or cream band around the back of the head, often surrounded by black spots. It has smooth scales on the back, and bands of rough scales on the sides to help it grip when climbing. It is light brown to grey above and pale-grey below, sometimes with dark flecks. Females are larger than males. **Behaviour** This snake is a very agile climber and lives among large trees lining rivers and creeks. It is active mainly at night, hunting in the trees and sometimes descending to the ground to forage. It shelters by day beneath loose bark, in hollow trunks and branches of old and dead trees, remaining alert for any prey that may venture into its shelter, and sometimes basking in the sun on warm days. It is generally shy, but becomes very aggressive if cornered or provoked, raising its fore-body in a rigid S-shaped curve, flattening its head and striking repeatedly at anything within range, with its mouth wide open. **Development** Females breed every second year. They mate in spring, and bear litters of 2–11 (usually 5–6) live young in late summer. The newborn are about 260 mm long. **Diet** A variety of vertebrates, especially frogs (mainly tree frogs), some geckos, skinks and small mammals, including bats. **Habitat** Dry sclerophyll forests, woodlands, sometimes rainforests and wet sclerophyll forests, along the coast and ranges, especially close to flowing water. **Threats** Habitat destruction and fragmentation due to clearing, forestry practices and agriculture, resulting in the loss of old trees with suitable sheltering sites.

LENGTH **To 1 m**	MIDBODY SCALES **19–21 rows**	STATUS **Vulnerable in NSW.**

BLACK TIGER SNAKE

DANGEROUSLY VENOMOUS

Now considered to be the same species as *Notechis scutatus*, this large, robust, highly venomous snake has a wide, blunt head, barely distinct from its neck, small eyes with round pupils and a dark iris. It has semi-glossy scales, black or very dark brown above, sometimes with faint paler or darker cross-bands. The belly is pale to dark grey or almost black. Those living in cooler climates tend to be darker, enabling them to warm up quickly in sunlight. **Behaviour** Predominantly ground-dwelling, active by day and on warm evenings, it shelters in abandoned burrows, among tussock grasses, in dense vegetation, beneath rocks and fallen timber. It often basks in the sun among low grass tussocks, and sometimes climbs trees to bask or search for food. It is generally shy, but if cornered it raises its fore-body in a loose curve with its head slightly up, inflating and deflating its body while hissing loudly. If provoked it will strike out and bite strongly and quickly, injecting some of the most toxic venom of all the land snakes. If disturbed from above it flattens its body and keeps close to the ground, thrashes jerkily, sometimes making wild, inaccurate bites. Large adults are often very sluggish and do little when approached or handled. In the breeding season rival males often wrestle for the right to mate with a female. **Development** Mating takes place from autumn through spring, and females give birth to 6–110 (usually 10–20) live young in mid to late summer. The newborn are about 150 mm long, and are sexually mature at about 2 years. **Diet** Frogs, tadpoles, lizards, birds, small mammals and fish. On Bass Strait they live among mutton bird colonies, feasting on young chicks and starving for the rest of the year. King Island snakes are cannibalistic. **Habitat** Wet and dry sclerophyll forests, woodlands, shrublands, coastal heaths, coastal dunes and the margins of rocky creeks, on the ranges, slopes, coastal plains and offshore islands. Also in grazing lands and other highly degraded areas.

EASTERN TIGER SNAKE

DANGEROUSLY VENOMOUS

A large, robust, highly venomous snake with a broad, blunt head, only slightly distinct from its neck. It has small eyes with a round pupil and dark iris. The scales are semi-glossy and are very variable in colour from light grey to yellow, olive, brown, reddish or blackish-brown above, usually with narrow yellowish cross-bands, giving the snake a tiger-like appearance. The belly is cream, olive-green to grey, often darker on the throat. Those living in cold areas tend to be darker, enabling them to warm up quickly in the sun. **Behaviour** This snake is predominantly ground-dwelling, although in floods it climbs into bushes and low trees to reach bird nests. It is active by day and also at night in warm weather, entering animal burrows looking for prey, using constriction as well as venom to kill its victim. It shelters under rotting timber, in abandoned burrows, beneath rocks and in dense vegetation, usually in wet areas, rarely staying in the same place for more than 15 days. In winter it becomes inactive, retreating to burrows, hollow logs and tree stumps. Groups of as many as 26 juveniles have been found overwintering together. It is shy and tends to flee if disturbed, becoming aggressive only when provoked. If cornered it raises its head slightly, flattens its fore-body into a low, open curve, swings from side-to-side while inflating and deflating its body and hissing loudly. If provoked it will lunge forward in various directions, striking when it can. Its venom is among the most toxic of all the land snakes. **Development** Mating takes place from autumn to spring, and in summer in Tasmania (reproducing every second year). Females give birth to 10–126 (usually 15–30) live young from January to April. The newborn are around 215 mm long with distinct thick brown and yellow bands, and reach sexual maturity at about 2 years of age. **Diet** Mainly frogs, with some lizards, birds, small mammals and fish. **Habitat** Rainforests to wet and dry sclerophyll forests, woodlands and heathlands, along the coast, ranges and western slopes, from sea level to above 1000 m. It prefers swamps and river flood plains, particularly in hot and dry regions, and is common in some suburban areas.

LENGTH **To 2 m** MIDBODY SCALES **17 or 19 (rarely 15) rows** STATUS **Low risk.**

INLAND OR WESTERN TAIPAN
FIERCE SNAKE

DANGEROUSLY VENOMOUS

A large, moderately robust, highly venomous snake with a long, narrow head slightly distinct from its neck. It has a round snout, a prominent brow ridge and long fangs. The pupils are round with a dark iris. The scales are smooth and glossy, varying from pale yellowish-grey, to olive-brown or rich brown above. Many scales have black edges, forming thin diagonal lines. The head and neck are usually glossy black in winter, fading to brown in summer, with scattered darker flecks. The belly is cream to yellow, sometimes with orange blotches. Males are the larger sex. **Behaviour** This ground-dwelling snake is most active in the early morning when it basks in the sun and searches for prey around cracks and crevices, detecting prey by sight and odour. In cool weather it hunts mostly in the afternoon, and on hot days is active after dark. This snake produces large quantities of highly toxic venom (the most potent of all the world's land snakes) and bites quickly several times without releasing its victim. It shelters in rat burrows (especially those of the long-haired rat), in deep soil cracks, sink holes and sometimes in rocky crevices. Shy, placid and very alert, it usually disappears into cover if a human approaches. If cornered or threatened it rears up, holds its fore-body in a low curve and will strike with great speed and accuracy. Rival males fight for the right to mate with a female, intertwining their bodies, wrestling and lunging with closed mouths. **Development** They breed from August to December and females lay 9–20 eggs, about 60 mm long, in a clutch. In good years they lay 2 clutches and in dry years they do not breed. The eggs hatch after about 10 weeks and the hatchlings are about 470 mm long. **Diet** Small to medium-sized mammals, especially the long-haired rat and plains rat. **Habitat** Found in arid to semi-arid drainage regions of inland river systems, in open shrublands and open herblands on flood plains with deep cracking soil, gibber plains, sand dunes and rocky outcrops.

LENGTH **To 2.7 m**	MIDBODY SCALES **23 (rarely 25) rows**	STATUS **Low risk.**

195

| FAMILY **ELAPIDAE** | SPECIES *Oxyuranus scutellatus* |

COASTAL TAIPAN

DANGEROUSLY VENOMOUS

A large, slender, highly venomous snake with a long, narrow head quite distinct from its neck. It has an angular brow ridge and long fangs. The eyes are large with a round pupil and an orange-brown iris in light-coloured and young individuals. The scales are glossy and paler in summer than winter, light olive to yellow-brown, reddish-brown, dark brown or almost black above in north Qld. The head is paler and the snout is cream-coloured. The belly is cream with scattered orange spots. Males are much larger than females. **Behaviour** Semi-nomadic and generally solitary, this snake shelters in abandoned burrows, hollow logs and stumps, and beneath deep surface litter. It usually basks and forages in the mid morning, although on cool days it emerges in the afternoon, and in hot weather is active at night. The size of this snake and the potency of its venom make it the most deadly of all the world's land snakes. It detects prey by sensing movement and odour and captures larger prey by striking rapidly and releasing quickly to avoid retaliation and potential injury, following its victim until it succumbs to the venom. It is shy, swift and very alert with keen eyesight, and usually disappears into cover if a human approaches. If cornered or threatened it holds its body in loose coils with its head and fore-body raised slightly, and will deliver one or more stabbing bites with great accuracy. Rival males engage in wrestling matches in the breeding season to determine their mating rights.

Development They breed from July to December, and females lay 3–21 soft-shelled eggs, about 55 mm long, from October to February, producing 2 clutches in a good season. The eggs hatch 60–80 days later and the hatchlings stay in the split egg cases for 1–2 days before emerging completely. They are about 44 mm long and grow rapidly. Males are sexually mature by 16 months, females by 28 months. **Diet** Small to medium-sized mammals (especially rodents and bandicoots) and birds. **Habitat** Dry sclerophyll forests and woodlands, monsoon forests, coastal heaths and cultivated areas, favouring dense thickets and grassy slopes.

| LENGTH **To 2.9 m** | MIDBODY SCALES **21–23 rows** | STATUS **Low risk.** |

FAMILY **ELAPIDAE**   SPECIES *Pseudechis australis*

MULGA OR KING BROWN SNAKE

DANGEROUSLY VENOMOUS

A large, robust, highly venomous snake with a broad, flattened head just distinct from its neck. It has small eyes with a round pupil and a reddish-brown iris. The scales are smooth, reddish-brown, coppery-brown to blackish-brown above, often with darker edges, forming a net-like pattern. The belly is cream to white, sometimes with orange blotches. Males are larger than females. **Behaviour** This ground-dwelling snake is active by day in cool weather, and during the evening and at night in hot weather. It shelters in abandoned animal burrows, below fallen timber, under large rocks, in deep rock crevices, and in deep soil cracks. It spends most of its day close to its shelter, and travels along familiar routes looking for food. If disturbed it is not highly aggressive and prefers to flee if possible. If cornered or harassed it flattens its neck and fore-body into a long, horizontal curve pointing to one side of the aggressor, hisses loudly, and may strike towards the opposite side, biting a number of times, injecting more venom than any other Australian snake. Males travel widely in the breeding season looking for females, and rival males wrestle with each other to determine mating rights. **Development** Females mate in spring and lay 4–19 soft-shelled eggs, about 50 mm long, in a sticky mass from December to February. They hatch 64–97 days later and the hatchlings are about 300 mm long. They have a lifespan of around 10 years. **Diet** A wide variety of vertebrate prey including skinks, dragons, snakes, frogs, birds, eggs, mice and other mammals. Invertebrates and carrion are sometimes eaten. **Habitat** Most habitats from monsoon forests to hummock grasslands, shrublands, gibber and sandy deserts. It is often found in dry woodlands, and occurs in highly disturbed areas such as wheatfields.

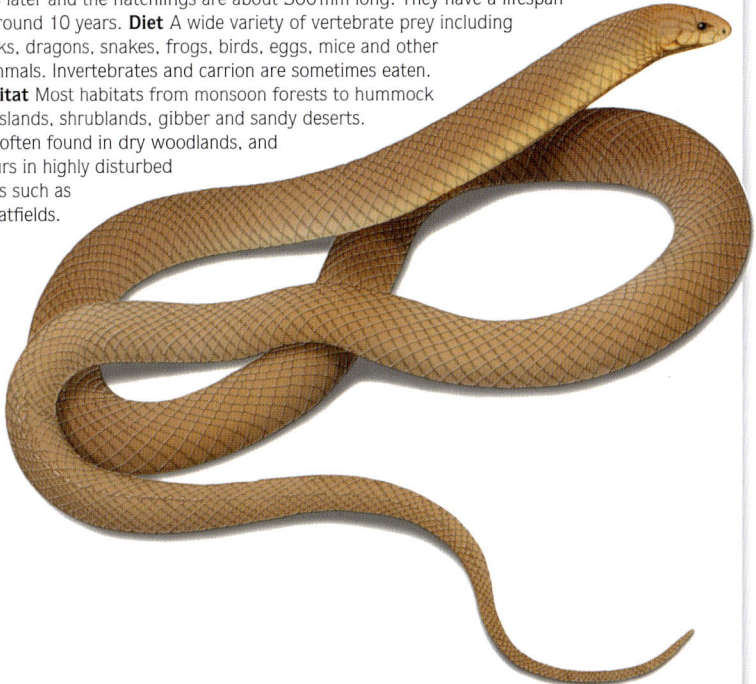

| LENGTH **To 3.3 m** | MIDBODY SCALES **17 rows** | STATUS **Low risk.** |

| FAMILY **ELAPIDAE** | SPECIES *Pseudechis guttatus* |

BLUE-BELLIED BLACK SNAKE
SPOTTED BLACK SNAKE

DANGEROUSLY VENOMOUS

A moderately large, robust snake with a broad, dark, flattened head just distinct from its neck, small eyes with a round pupil and a black iris. It has smooth scales, usually shiny bluish-black to brownish-black above, varying through shades of grey to pale brown, with various amounts of paler grey, brown or cream spotting. Some individuals are cream-coloured. The belly is dark bluish-grey to bluish-black. Males grow larger than females. **Behaviour** This ground-dwelling snake is most abundant from August to November, and is active mainly in the mornings and afternoons, and at night in hot weather. It often basks in the sun and shelters beneath fallen timber, in abandoned animal burrows, among rocks, in deep soil cracks, in dense vegetation and in rubbish dumps. This snake is usually shy, stays close to its shelter and generally retreats when discovered. If provoked it can become very aggressive, raising itself slightly from the ground, flattening its neck, arching its fore-body, and hissing loudly. Before attempting to bite, it swings from side-to-side and often strikes with its mouth closed. The venom is very toxic but no human deaths have been recorded. In the breeding season males engage in wrestling matches, coiling tightly around each other and rolling around. Male combat has been observed in late winter and early spring and involves biting, intertwining for about 15 minutes and then pursuit with more biting. **Development** They breed in late winter and early spring, and females lay 6–16 soft-shelled eggs in December. The eggs hatch 44–87 days later. The hatchlings are about 300 mm long and stay in their slit egg cases for 1–2 days consuming the remaining yolk. **Diet** Frogs, lizards, snakes, skinks and small mammals (especially rabbits, mice and rats), and the occasional invertebrate. **Habitat** Dry sclerophyll forests, woodlands and grasslands along the coastal plains, slopes and ranges, often around lakes and river flood plains.

| LENGTH **To 1.9 m** | MIDBODY SCALES **19 rows** | STATUS **Low risk.** |

FAMILY **ELAPIDAE** SPECIES *Pseudechis porphyriacus*

RED-BELLIED BLACK SNAKE

DANGEROUSLY VENOMOUS

A large, robust snake with a broad, flattened head just distinct from its neck. It has small eyes with a black iris and round pupil. The scales are smooth and iridescent purplish-black above. The snout is often pale brown. The lower sides are bright red to deep pink, fading to dull pink or cream on the belly. The underside of the tail is black. Males are larger than females. **Behaviour** Frequently encountered near water, this snake is active during the day and on warm evenings and nights, and hibernates in cold weather. It shelters and hunts in a home range of around 1 ha, actively searching for prey in small crevices and in the water. It uses a number of shelter sites, including abandoned burrows, hollow logs, large flat rocks and dense vegetation. In the spring breeding season males spend a long time in the open and on the move, travelling up to 1220 m in a single day searching for mates within an expanded home range of around 70 ha. Rival males wrestle with each other to determine mating rights, intertwining their bodies and attempting to force the opponent's head down. Small groups of pregnant females may share a burrow and often bask together by a creek. If disturbed it moves away quickly, but if cornered or provoked it becomes aggressive, raises its fore-body off the ground with its neck flattened and hisses loudly. It is reluctant to bite, and only one human has reportedly died after being bitten. **Development** They mate in spring and females give birth to 5–18 live young, about 280 mm long, from January to March. The newborn emerge from membranous sacs shortly after birth. They become sexually mature between 19 and 31 months. **Diet** A variety of vertebrates, mostly frogs, tadpoles and lizards. Also some other reptiles, mammals and fish, especially eels. Poisonous cane toads are also eaten, causing declines in the snakes' population in some areas. **Habitat** Wet and dry sclerophyll forests, woodlands, rainforests, shrublands and grasslands along the slopes, ranges and lowlands, often around waterways.

| LENGTH **To 2.54 m** | MIDBODY SCALES **17 rows** | STATUS **Low risk.** |

| FAMILY **ELAPIDAE** | SPECIES *Pseudonaja affinis* |

DUGITE
SPOTTED BROWN SNAKE

DANGEROUSLY VENOMOUS

A slender snake with a narrow head indistinct from its neck, an angular brow and large eyes with a round pupil encircled by a narrow golden ring. It has smooth, glossy scales, grey, olive or brown above, with scattered black spots sometimes covering most of the back. The lining of the mouth is bluish-black, and the head is often paler than the body. The belly is pale grey or pale brown with orange or brownish blotches. This snake can change colour, becoming darker over the course of a few months. **Behaviour** This fast-moving, ground-dwelling snake has very acute vision, and subdues its prey using both constriction and venom. It is active during the day and on warm evenings and nights searching for prey, looking in their shelter sites and under surface debris, constricting as well as biting its victim. It shelters beneath logs, rocks, surface debris, in hollow logs, abandoned burrows, termite mounds and stick ant nests. It often overwinters in a burrow. It is shy and usually avoids encounters with humans, but is easily agitated. If confronted or provoked it raises its body into a double curve, hisses loudly and will strike fast and accurately, usually high up on the leg or arm. Males roam around in the breeding season looking for females and wrestle with rivals to determine the right to mate with a particular female. **Development** They mate in late winter and early spring. Females lay 11–35 eggs, about 32 mm long, in a deep rocky recess or in a disused animal burrow in late spring to mid-summer. The eggs hatch 52–105 days later. The hatchlings are about 210 mm long and have a lifespan of 16 years or more. **Diet** A variety of vertebrates, especially house mice, birds, lizards, frogs and sometimes other snakes, including its own species. **Habitat** Sclerophyll forests, woodlands, shrublands, heathlands and coastal dunes on the plains, slopes and ranges, including arid areas. It is also found in degraded habitats, including golf courses, industrial parks and gardens.

| LENGTH **To 2.13 m** | MIDBODY SCALES **19 rows** | STATUS **Low risk.** |

SPECIES *Pseudonaja modesta*

RINGED BROWN SNAKE
VENOMOUS BUT NOT DANGEROUS

A small, relatively slender snake with a narrow head indistinct from its neck. It is distinguished by its black head separated by a narrow cream-coloured bar from a broad black band at the back of the neck, and 4–12 evenly-spaced, narrow, dark bands between the neck and the tip of its tail. It has an angular brow, large eyes with a round pupil surrounded by a thin reddish-orange rim, and a black iris. The mouth is bluish-black inside with short fangs. The scales are smooth and glossy, pale grey or brown to rich reddish-brown above. The belly is cream to pale yellow with orange blotches. The head markings and bands fade with age. **Behaviour** This is a fast-moving, ground-dwelling snake with keen eyesight. It is active by day and on evenings and nights in hot weather, searching for prey which is subdued by constriction as well as venom. It shelters in abandoned lizard burrows, under fallen timber, beneath surface debris and in clumps of spinifex grass. It is a competent burrower and sometimes digs a burrow in soft sandy soil, using its head to flick the sand out sideways. It is very shy and avoids confrontation, but if threatened or provoked it raises its fore-body into an S-shaped curve and strikes rapidly. Although the venom is very toxic, the small amount injected is not enough kill a human. **Development** They mate in spring and females lay 2–10 eggs. The hatchlings are about 130 mm long. **Diet** Mainly skinks and geckos, with some small mammals. **Habitat** Woodlands, shrublands, mallee and hummock grasslands in semi-arid to arid areas, avoiding black-soil plains. **Threats** Loss of habitat due to land clearing, habitat fragmentation and degradation due to grazing by stock, predation by foxes and cats, hunting and illegal collection.

| LENGTH **To 600 mm** | MIDBODY SCALES **17 rows** | STATUS **Endangered in NSW.** |

| FAMILY **ELAPIDAE** | SPECIES *Pseudonaja mengdeni* |

WESTERN BROWN SNAKE
GWARDAR

DANGEROUSLY VENOMOUS

A large, slender snake with a narrow head indistinct from its neck, an angular brow, and large eyes with a round pupil and a pale iris. It has short fangs and a bluish-black mouth lining. The scales are smooth, glossy, and vary greatly in colour, from light brown to orange-brown, olive-brown or almost black above. Some individuals have various patterns including V- or W-shaped black marks on the neck, broad dark cross-bands, narrow brown cross-bands and herring-bone patterns. The belly is cream to yellow with dark grey or dark orange blotches. They become paler in spring and summer. Males are larger than females. **Behaviour** This fast-moving snake is predominantly ground-dwelling although it occasionally climbs trees. It is active by day and on evenings and nights in hot weather. It has acute vision and searches for prey which it constricts while its venom takes effect. It shelters under logs, surface debris, below large flat rocks, in deep soil cracks, in hollow logs and abandoned mammal burrows. Shy and quite nervous, it is reluctant to bite and tries to avoid humans, but if cornered or provoked it raises its body into an S-shape, flattens its neck, hisses loudly and may strike rapidly before attempting to escape. Rival males compete in wrestling matches in the breeding season to determine which will mate with a particular female. **Development** They mate in spring and early summer in the south, and females lay 7–38 eggs, about 35 mm long, from November to February, producing 2 clutches in good seasons. Northern females are less seasonal in their breeding. The eggs hatch 61–93 days later, and the hatchlings are about 220 mm long. **Diet** Small mammals (especially house mice), skinks, geckos, dragons, reptile eggs, occasionally frogs and smaller snakes.

Habitat Sclerophyll forests and woodlands, arid scrubs, spinifex dune fields and sandplains in arid and semi-arid areas, avoiding black-soil plains. It is also found in disturbed areas around cities, towns and farmlands.

| LENGTH **To 1.6 m** | MIDBODY SCALES **17–19 rows** | STATUS **Low risk.** |

FAMILY **ELAPIDAE**   SPECIES *Pseudonaja textilis*

EASTERN BROWN SNAKE
DANGEROUSLY VENOMOUS

This slender snake has a small, narrow head indistinct from
its neck, a prominent brow and large eyes with a black
iris and a round pupil partly encircled by an orange or pale
brown rim. Unlike the similar western brown snake it has
a pink mouth lining. The scales are smooth and semi-glossy,
and vary from greyish-yellow through brown, reddish-brown
and orange to almost black above. Adults usually lack patterning,
while juveniles usually have a black head, a black band on the back of the
neck and numerous black cross-bands. The belly is cream, yellowish or orange
with darker orange or dark grey blotches. Males are larger than females.
Behaviour Swift and alert, it has acute vision and actively hunts for prey,
biting and constricting victims. It is active by day and by night in hot weather,
sheltering under logs, surface debris, in rock crevices, deep soil cracks, hollow
logs and abandoned burrows. It hibernates in a burrow in winter, sometimes
in large groups. It is very nervous and often becomes aggressive if disturbed,
although it generally tries to escape, and will usually tolerate being accidentally
trodden on when concealed. If cornered or provoked it hisses loudly, flattens
its neck, raises its fore-body into an upright double S-shape, and strikes very
quickly, usually high up on the leg or arm, occasionally chasing its aggressor
and striking repeatedly. The venom is very toxic, but only small amounts are
injected through its short fangs. Rival males wrestle for mating opportunities
in the breeding season. **Development** They breed in spring and early summer.
Females lay a clutch of 6–25 soft-shelled eggs, about 36 mm long, from October
to December, sometimes in a communal nest. They hatch 36–75 days later.
The hatchlings have bands on the head and neck, and are about 240 mm long.
In good seasons females lay 2 clutches. **Diet** Small mammals (especially mice)
and reptiles, with some frogs, birds and reptile eggs. **Habitat** Open forests,
woodlands, heathlands, shrublands, mallee, savannah grasslands,
around inland watercourses, swamps and seasonally inundated
areas. It is also found in disturbed sites including farmlands,
semi-rural and rural areas.

Juvenile

LENGTH **To 2.2 m**   MIDBODY SCALES **17 rows**   STATUS **Low risk.**

| FAMILY **ELAPIDAE** | SPECIES *Simoselaps bertholdi* |

Desert Banded Snake

VENOMOUS BUT NOT DANGEROUS

A small, burrowing snake distinguished by its black and yellow bands. It has a small head indistinct from its neck, a flattened rounded snout, a short tail, and small eyes with a pale iris and vertical pupil. It has smooth, glossy scales and an enlarged scale on the tip of the snout to protect it when burrowing. The head is grey-brown or black with a whitish snout mottled with black, brown and grey. The body is cream, yellowish or reddish-orange, with 18–31 black rings of about equal width encircling the body and tail, and extending onto the creamy-white belly. Females are larger than males. **Behaviour** This snake burrows into the loose upper layers of sandy soil and seems to 'swim' through the sand. It chooses sheltered sites beneath overhanging vegetation, and also below embedded fallen timber and stumps. It ambushes passing prey, lying partially buried with its head protruding above the sand or ground litter. It is predominantly nocturnal, but warms itself in the loose surface sand on cool sunny days. In hot weather it is sometimes seen on the surface in the day, particularly in spring when males are searching for mates. Prey is killed by the injection of venom, and held tightly in the snake's coiled body until the venom takes effect. This snake is very shy and does not attempt to bite humans. **Development** They mate in spring and females lay 1–8 soft-shelled eggs in late summer. The hatchlings are about 70 mm long. Males are sexually mature at about 20 months, and females at 32 months. **Diet** Skinks and legless lizards. **Habitat** Woodlands, shrublands and heathlands in sand dune deserts, sand plains and rocky outcrops.

| LENGTH **To 330 mm** | MIDBODY SCALES **15 rows** | STATUS **Low risk.** |

FAMILY **ELAPIDAE** SPECIES *Suta suta*

CURL OR MYALL SNAKE

VENOMOUS — LARGE INDIVIDUALS MAY BE DANGEROUS

A fairly stout snake with a broad, flattened head distinct
from its neck. The eyes have a vertical pupil and a pale iris.
It is smooth, pale fawn to reddish-brown above with paler
flanks and sometimes with a narrow black stripe along the
back. The scales may have a dark brown edge, forming a
net-like pattern over the back. The head and neck are dark
brown to black with a paler stripe above and through the eye to
the temple, and white or cream upper lips. The belly is creamy-white. Males are
larger than females. **Behaviour** This ground-dwelling snake is active at night
and is often seen foraging in open areas on warm nights. In addition to moving
by lateral undulation, this snake can also move by side-winding. It shelters
during the day in deep soil cracks, beneath fallen timber, under rocks, in
disused animal burrows, in grass hummocks and hidden within deep leaf litter.
If threatened or provoked it flattens its body, curls itself into a tight, spring-like
coil, thrashes wildly and strikes with fast, snapping bites. **Development** They
mate from autumn to spring, and females give birth to 1–7 live young, about
160 mm long, in late spring and summer. **Diet** A variety of vertebrate prey,
mainly lizards and small mammals, with some frogs and snakes. **Habitat** Sub-
humid to arid tropical woodlands, shrublands and hummock grasslands, in the
plains, rocky ranges and slopes. Prefers deeply cracking clay soils, and lives
along dry watercourses in desert areas.

| LENGTH **To 880 mm** | MIDBODY SCALES **19 (rarely 21) rows** | STATUS **Low risk.** |

FAMILY **ELAPIDAE**   SPECIES *Tropidechis carinatus*

ROUGH-SCALED OR CLARENCE RIVER SNAKE

DANGEROUSLY VENOMOUS

A fairly robust snake with a broad, glossy, triangular head distinct from its neck. It has a prominent brow ridge, round pupils and a dark iris. The scales are rough with projecting ridges. They are matt, yellowish-brown, greenish-brown to dark brown or almost black above, usually with narrow, dark cross-bands that become less distinct on the tail and in older individuals. The belly is cream, olive or grey, usually with dark blotches. **Behaviour** The rough-scaled snake is both ground-dwelling and arboreal, climbing to 5 m in trees and shrubs. It is generally active at night and around dawn and dusk, but may bask or forage during the day in cooler, sunny weather. It shelters in tree hollows and among dense foliage, sometimes quite high in a tree, often curling up in the centre of a large epiphytic fern. On the ground it shelters under fallen timber, in soil cracks and cavities in the roots of strangler figs. It climbs into the low foliage to hunt or bask in the sun, staying close to its shelter and descending to the ground late in the afternoon to ambush frogs. Shy and nervous, it prefers to escape rather than confront a human, although it becomes very aggressive when cornered or provoked, raising its neck and fore-body in an upright S-shape. It may hiss in short bursts and strike with several fast, accurate bites before trying to escape. **Development** They mate in spring and summer. Females give birth to 5–18 live young, about 200 mm long, in late summer and autumn. Females do not breed every year. **Diet** Mainly small mammals and frogs, with some birds and reptiles. **Habitat** Rainforests, sclerophyll forests, heathlands, shrublands, woodlands and disturbed habitats along the coast and ranges to 1100 m, usually close to swamps and creeks with dense vegetation.

LENGTH **To 1.2 m**	MIDBODY SCALES **23 rows**	STATUS **Low risk.**

FAMILY **ELAPIDAE**   SPECIES *Vermicella annulata*

BANDY-BANDY

VENOMOUS BUT NOT DANGEROUS

A medium-sized, slender, burrowing snake with a short, blunt tail and a narrow, flattened head indistinct from its neck. It has a round snout with a dark tip and very small eyes with a dark iris. Although venomous, it has a very small mouth making it very difficult to bite an adult human. There is a white band between the eye and the nostril. The scales are smooth and glossy, and form a distinctive pattern of bluish-black and white bands encircling the body. The number of bands ranges from 48–147. Females grow much larger than males. **Behaviour** This snake is most active at night and shelters by day in a burrow dug into loose soil under rocks or logs, in soil cracks and termite mounds, often using the same shelter site for months. It is very shy and secretive, and is usually only seen on the surface when it is forced out of its burrow by heavy rain, or on warm nights when it is looking for a mate, searching for prey, or moving to a new shelter. The snake has reduced eyes and uses scent to find prey, flicking the ground and the air with its tongue to detect and follow the scent trails of blind snakes. It has a very slow metabolic rate, and after feeding becomes inactive and may stay under cover for up to 3 months. Although venomous, it is reluctant to bite. When alarmed it flattens its body and contorts itself into one or more vertical loops, raising itself high above the ground, interspersed with bouts of thrashing around. **Development** They breed from late October to late December. Females only breed every few years, laying a clutch of 2–13 eggs, about 35 mm long, from January to March. The eggs hatch some 62–64 days later and the hatchlings are about 190 mm long. Males reach sexual maturity at 2 years, females at 3 years. **Diet** Blind snakes and possibly burrowing skinks. **Habitat** A wide range of habitats from rainforests and wet sclerophyll forests to savannah woodlands, mallee and mulga shrublands, and dry spinifex sandhills. **Threats** Land clearing, pesticide use, feral cats.

| LENGTH **To 750 mm** | MIDBODY SCALES **15 rows** | STATUS **Vulnerable in Vic.** |

| FAMILY **HYDROPHIIDAE** | SPECIES *Astrotia stokesii* |

STOKES' SEA SNAKE

DANGEROUSLY VENOMOUS

This is one of the world's largest and bulkiest sea snakes, and potentially one of the most dangerous. Its bulky, muscular head is distinct from its neck, and it has large fangs and large venom glands. The nostrils are on top of the snout and have valves that close when the snake is submerged. The skin is loose and the tail flattened and paddle-shaped. It has small eyes with round, black pupils surrounded by a pale yellowish rim and a black iris. It is creamy-white to dark grey or almost black above with a pattern of dark blotches on the back. The markings fade with age and young snakes may have strong black bands. The scales overlap strongly, and those on the belly are very long and divided in the middle to form a keel along the snake's belly. **Behaviour** Active by day and night, this snake forages in crevices and holes on the sea floor and among coral reefs for slow-moving fish. It is a powerful swimmer and can stay submerged for long periods. It is sometimes seen on the surface around the coast and in deeper reef waters, occasionally leaving the water at low tide. It is not usually aggressive, but if trapped or provoked it will bite savagely with a chewing motion, or with a series of rapid bites. The fangs are able to penetrate a wet suit. **Development** Females give birth while submerged to a litter of 4–20 live young, about 390 mm long, in mid to late summer. **Diet** Fish. **Habitat** Tropical coastal waters from 3–25 m deep, around coral reefs and sandy or muddy sea floor areas. It sometimes strays into temperate seas in late summer.

| LENGTH **To 1.6 m** | MIDBODY SCALES **46–63 rows** | STATUS **Low risk.** |

FAMILY **HYDROPHIIDAE**   SPECIES *Pelamis platurus*

YELLOW-BELLIED SEA SNAKE

DANGEROUSLY VENOMOUS

A moderately-sized sea snake with a large narrow head distinct from its narrow neck, and a flattened, paddle-shaped tail. A wide black to dark bluish-brown stripe runs along the spine, contrasting with the cream, yellow or pale brown sides and belly. The tail is yellow with black bars or spots. The head is bluish-black above with yellow lips and chin. The nostrils have valves allowing it to submerge.

Behaviour This snake drifts around the oceans, carried by winds and currents. It spends most of its time below the surface, staying submerged for more than 3 hours at times, absorbing oxygen from the water through its skin. It feeds among surface rafts of seaweed and floating debris where small fish shelter. Thousands of snakes may be found feeding in a large slick. It waits for fish to come close, or slowly approaches its victim, suddenly flicking its head to the side or to the rear, with its mouth wide open, to snatch a small fish. Its body colouration probably warns aquatic animals of its poisonous nature, saving it from attack by predators. If approached by a swimmer or boat it usually dives to avoid confrontation, and is capable of swimming backwards or forwards equally well. Beached snakes and those in shallow coastal waters are usually listless, but if approached they may bite, and have an extremely toxic venom. It removes old skin and cleans itself by coiling into a knot and slowly swimming through the knot which breaks open at the tip of the tail. **Development** They breed in the warmer months in cooler waters, and probably year round in warm waters. Females often congregate in birthing areas and give birth to 1–7 live young, about 250 mm long, born after 5–6 months gestation. The young grow quickly and are sexually mature at 1–2 years. **Diet** Mainly surface fish. **Habitat** Open seas, with temperatures above 20ºC, usually more than 100 m deep, above continental shelves, around islands and sometimes in estuaries and coastal waters not diluted with fresh water.

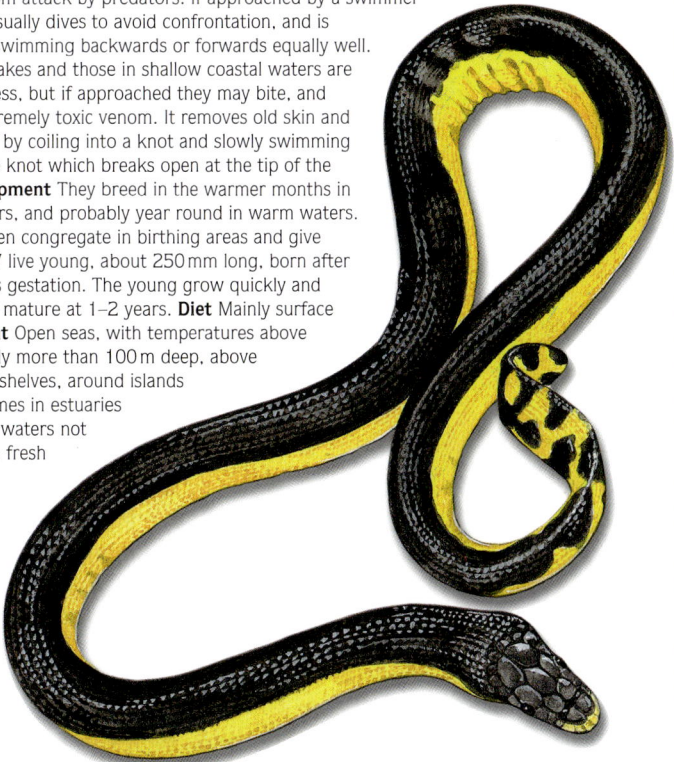

LENGTH **To 1.13 m**   MIDBODY SCALES **47–69 rows**   STATUS **Low risk.**

FRESHWATER CROCODILE

A fierce, amphibious reptile with thick horny skin over bony body armour. Valves in the nostrils and throat keep water out while submerged and its powerful tail and webbed feet propel it through the water. The jaws can crush turtle shells with ease, and have very sharp intermeshing teeth. The limbs are short and strong. It is grey to olive brown above with darker markings sometimes forming bands across the back and sides. Males are larger than females. Unlike the estuarine crocodile it has a long, slender and smooth snout.

Behaviour Generally shy and secretive, this crocodile slips silently under water at the slightest disturbance, remaining submerged for up to 1 hour. It often basks by day in shallow water or on riverside banks, rocks and logs. It usually forages on warm evenings and nights, and is easily spotted by its bright red eye-shine. It has acute senses of sight, hearing and smell, and sensory pits in the snout detect water movements made by nearby animals. It lies still in shallow water and ambushes small animals in the water or at the edge, snapping sideways at them. Larger prey is sometimes stalked. It swims fast with its legs held against its body, and can gallop at up to 18 kph over short distances on land. Humans are rarely threatened, although females guarding hatchlings may be aggressive until they disperse in the wet season floods. Adults often congregate in the dry season and may travel long distances overland to familiar waterholes. Some enter a state of torpor for up to 4 months, buried in the mud. Dominant males defend the centre of large pools, attacking and biting the tails of intruders. **Development** They mate early in the dry season and females lay 4–21 hard-shelled eggs (about 60 mm long) some 6 weeks later. They are laid in a moist hole, 60–450 mm deep, dug into a sandbank, usually within 10 m of the water and back-filled. The eggs hatch 2–3 months later, usually before the first wet season floods. Embryos developing slowly (at low temperatures) or quickly (at high temperatures) tend to be female, the remainder are generally male. The hatchlings are about 250 mm long and squeak to attract a female, not necessarily the mother, who digs them out and may carry them to the water in her mouth. The young gather in a crèche and often stay with the female for months. Maturity is reached at 9–17 years. **Diet** Insects, fish, frogs, crustaceans, small reptiles, birds, mammals and carrion. **Habitat** Permanent freshwater rivers, swamps and billabongs with fringing vegetation or rocky overhangs and banks, and some tidal estuaries where estuarine crocodiles are absent.

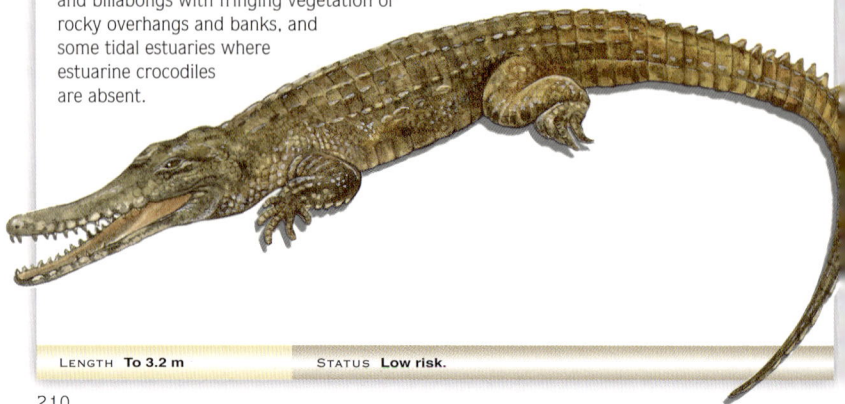

LENGTH **To 3.2 m**   STATUS **Low risk.**

ESTUARINE CROCODILE

The world's largest and most dangerous reptile, it is similar in most respects to the freshwater crocodile from which it can be distinguished by its shorter, broader, granular snout and heavier body armour. This powerful predator is able to crush the bones of mammals as large as a buffalo in its jaws. It is grey or brown to almost black above with irregular dark markings, and whitish below. Males are larger than females.

Behaviour Generally lethargic, this crocodile usually basks by day and hunts at night. It moves stealthily, close to the water's edge, with only its nostrils above water, resembling a floating log. It strikes with surprising speed, driving itself out of the water with its powerful tail, jaws agape, to snatch its prey. Large animals are dragged into deep water and rolled over until they drown, and then dismembered. Salty tears are produced to rid the body of excess salt ingested with the food. The dry season is spent mainly in estuaries, and adults congregate upstream in tidal rivers at the end of the dry and move to freshwater swamps where they spend the wet season. Females are very aggressive during the nesting season and guard their hatchlings for 2–3 months. Large adults force the juveniles to disperse either upstream or out to sea to find another estuary. Males are territorial and rivals engage in noisy and often ferocious combat at the onset of the breeding season. **Development** They breed at the beginning of the wet season. Females lay 30–70 hard-shelled eggs some 80 mm long, usually at night, inside a mound (about 450 mm high and 2 m across) of plant matter and mud, within 10 m of the water. The female guards the nest until the eggs hatch about 3 months later, when she digs out the chirping hatchlings and gently rolls any intact eggs around in her mouth to help them hatch. She carries the hatchlings in her mouth to the water where they gather in a crèche. Males reach maturity at around 16 years, females at 12 years.

Diet Any small or large animal, including fish, crustaceans, birds, frogs, bats, wallabies, dingoes, rodents and carrion. Hatchlings feed mainly on crabs, prawns and insects. **Habitat** Tropical estuaries, swamps and billabongs in the dry season, moving to freshwater swamps in the wet season.

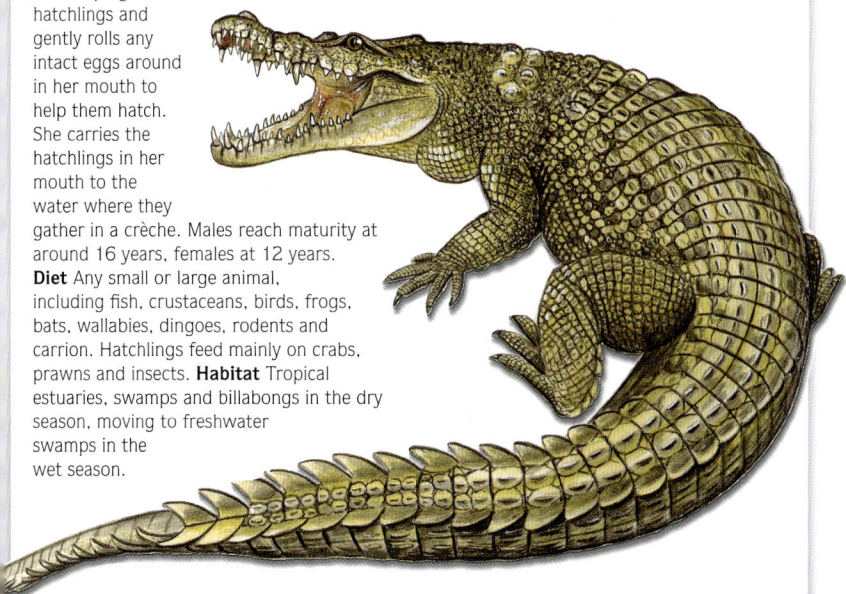

LENGTH **To 6.2 m**　　　STATUS **Low risk.**

FAMILY **CHELONIIDAE**   SPECIES *Caretta caretta*

LOGGERHEAD TURTLE

A very large sea turtle with strong, paddle-shaped limbs without webbing, a single claw on each flipper, and no obvious ankle joints. Older adults have a massive head and a heart-shaped shell with 5 plates (costal shields) on each side. It is dark reddish-brown above, sometimes with darker speckles, and white to yellowish below. Hatchlings are rich reddish-brown above and dark brown below. Like the other sea turtles it is unable to withdraw into its shell for protection, and has rough scaly skin on the head and neck. **Behaviour** This carnivorous turtle forages in deep water around coral reefs and bays, to about 50 m offshore, and is able to crush mollusc shells with its powerful jaws. It can stay submerged for up to 30 minutes, using its front flippers for propulsion and its smaller hind flippers for steering. Young turtles are dispersed by ocean currents and are typically found drifting on floating mats of sargassum weed in warm ocean currents. They ride these for several years until they are about 300–400 mm long when they settle in a feeding area and remain tied to it for the rest of their lives. At intervals of 1–10 years, mature turtles migrate from their feeding grounds back to the waters around their original nesting beach to mate, which may be up to 2600 km away. Females mate underwater, often with several males. The males stay offshore while the females alternate between mating in the water and nesting on land. **Development** Females lay approximately 3 clutches of 125 parchment-shelled eggs, about 40 mm diameter, at 2 week intervals between October and February, mating between each clutch. The eggs are laid in the sand above high tide in an egg chamber about 400–700 mm deep at the bottom of a large depression dug with the flippers and covered with sand. The eggs hatch 45–70 days later. The proportion of males to females depends on the incubation temperature. The 30–50 mm long hatchlings dash to the water at night and swim out to sea for several days. Sexual maturity is reached between 10 and 30 years, and they may live to 62 years or more. **Diet** Fish, molluscs, crustaceans, sponges and jellyfish. Young turtles feed mostly on plankton, algae and molluscs.

Habitat Tropical and warm temperate waters worldwide, and occasionally in southern Australian waters. They nest on the mainland and islands around the Tropic of Capricorn.

Threats Drowning in prawn-trawling nets, egg collection and disturbance of nesting sites, pollution, beachfront development.

SHELL LENGTH **To 1.5 m**   STATUS **Endangered.**

FAMILY **CHELONIIDAE**   SPECIES *Chelonia mydas*

GREEN TURTLE

A large sea turtle with a relatively small, beaked head and large scales on the upper eyelids. It has strong, paddle-shaped limbs without webbing and no obvious ankle joints. The shell is almost circular to heart-shaped with 4 plates (costal shields) on each side. It is yellowish to olive-green above, usually with reddish-brown to black markings, and whitish or cream below. Hatchlings are shiny black above with white-edged flippers and white underparts. Males are larger than females. **Behaviour** This turtle spends most of its life at sea, resting in rock holes, under rock overhangs and in underwater caves, surfacing every 30 minutes or so to breathe (adults can remain submerged for up to 5 hours). They also sometimes bask in the sun on secluded beaches. Young turtles travel around with the ocean currents until they are about 300–400 mm long, at 3–5 years old, living on floating mats of sargassum weed, feeding on small crustaceans and other sea creatures. They then settle in a shallow-water feeding area for the rest of their lives. Mature adults return to waters around their original nesting beach to breed every few years, which may involve a journey of thousands of kilometres. Females mate with a number of males on the surface or underwater. They store the sperm for use later in the season and return to land to lay their eggs. **Development** They mate at any time of year with a peak in summer, and females lay about 110 parchment-shelled eggs, about 44 mm diameter, on islands and coastal beaches. The eggs are buried in sand above the high water mark after dark, in a nest chamber 400–700 mm deep. An average of 5 clutches are laid in a season at intervals of 10–15 days, in separate nests on the same beach, followed by a break of 1–9 years. The eggs hatch 40–72 days later, and the proportion of males to females depends on the incubation temperature. The hatchlings are about 50 mm long and dash to the water at night where they swim and drift to the open ocean. Adults become sexually mature between 9 and 24 years and may live to 50 years.

Diet Adults are strictly herbivorous, eating seagrass, seaweed and algae. Juveniles also eat jellyfish, crabs, sponges, snails and worms.

Habitat Shallow tropical and warm temperate coastal waters worldwide. **Threats** Unsustainable commercial harvesting, coastal development, drowning in prawn-trawling nets, predation of eggs by feral animals.

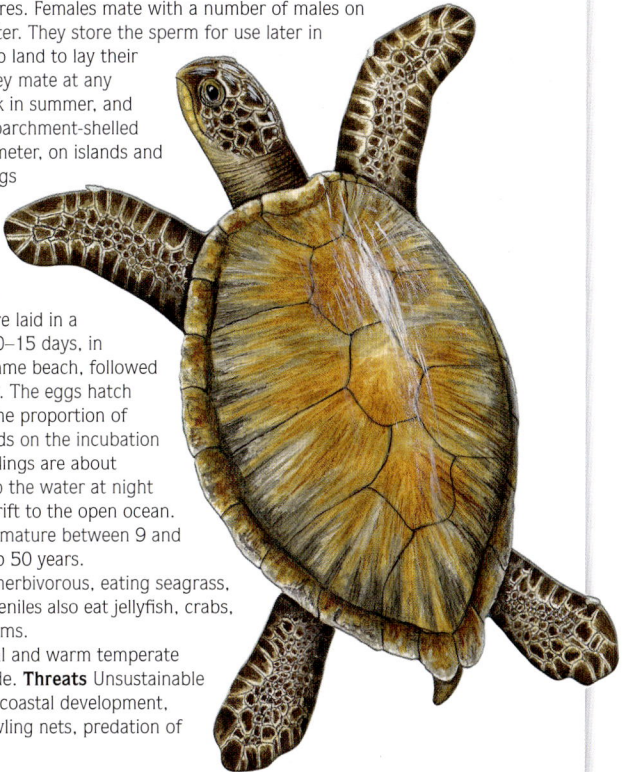

SHELL LENGTH **To 1.5 m.**   STATUS **Endangered.**

| FAMILY **CHELONIIDAE** | SPECIES *Eretmochelys imbricata* |

HAWKSBILL TURTLE

A sea turtle with a relatively small head and a distinctive beak-like upper jaw. It has strong, paddle-shaped limbs without webbing and no obvious ankle joints. Each forelimb has 2 claws. The shell is narrowly heart-shaped and serrated at the rear. It has 4 plates (costal shields) on each side, and in adults the plates overlap. It is olive-green or brown above with patches of reddish-brown and dark brown or black, and cream below. **Behaviour** This carnivorous turtle forages around coral reefs, using its beaked jaw to prise molluscs, crustaceans and other marine creatures from crevices in rocks and coral. They are usually seen resting by day in caves and large crevices. Hatchlings are dispersed by ocean currents and live on floating mats of sargassum weed for several years until they are 300–400 mm long. At this stage they settle in a shallow water feeding area and remain tied to it for the rest of their lives. Mature adults return to the waters around their nesting beach every few years to breed, mating in shallow lagoons. Some travel up to 2400 km while others feed and nest in the same area. In Australia they nest mainly on Great Barrier Reef and Torres Strait islands. **Development** Females lay clutches of about 120 round, parchment-shelled eggs in a nest chamber 400–700 mm deep dug into sand above the high water mark. They usually lay 2–4 clutches at intervals of 2 weeks in separate nests on the same beach. Eggs are usually laid in summer at night, followed by an interval of 2–4 years before breeding again. The eggs hatch 52–57 days later and the proportion of males to females is determined by their incubation temperature. The hatchlings are about 25 mm long and dash to the water under cover of darkness. They swim directly out to sea for several days. **Diet** Mainly sponges, some fish, crustaceans, molluscs, anemones, marine algae, jellyfish. **Habitat** Tropical and warm temperate coastal waters worldwide, especially around coral reefs. **Threats** Hunting, marine debris and pollution, coastal development, predation of eggs.

| SHELL LENGTH **To 1m** | STATUS **Critically endangered.** |

FLATBACK TURTLE

A large sea turtle with a low-domed, more or less heart-shaped shell with upturned edges. It has a pointed snout, strong, paddle-shaped limbs without webbing, a single claw on each flipper, and no obvious ankle joints. The shell has 4 plates (costal shields) on each side and in adults is covered by a thin fleshy skin. It is grey to pale grey-green or olive above, and the underside is creamy-yellow. Hatchlings are olive-green above with black-edged plates. **Behaviour** This carnivorous turtle spends its life at sea, propelling itself with its front flippers, and steering with its hind flippers. It floats on the surface when resting, and can stay submerged for up to 30 minutes. Young turtles stay in shallow waters off the coast and eventually settle in a shallow-water feeding area for the rest of their lives. This may be up to 1300 km from the nesting site. Mature turtles travel back to waters around their nesting beach every few years to mate at sea. Females nest on the Australian mainland and on continental islands, and tend to bite if approached. **Development** Females lay an average of 3 clutches of about 50 parchment-shelled eggs, about 52 mm diameter, between October and January in southern Qld, and mainly between June and August in northern Australia. The eggs are deposited in beach sand above the high water mark, in a nest-chamber dug about 400–700 mm deep. Clutches are laid in separate nests on the same beach, followed by a break of 18 months to 4 years. The eggs hatch after about 6 weeks, and the proportion of males to females is determined by the incubation temperature. The hatchlings are about 60 mm long, and dash to the water at night. They remain in shallow waters and take 30–50 years to reach maturity. **Diet** Brown algae, cuttlefish, soft corals, sea pens and sea cucumbers. **Habitat** Shallow tropical waters over the continental shelf of northern Australia. **Threats** Drowning in prawn-trawling nets, nest predation by feral pigs, disturbance by vehicles, beach development, pollution of waterways.

FAMILY **CHELIDAE** | SPECIES *Chelodina longicollis*

EASTERN LONG-NECKED TURTLE
SNAKE-NECKED TURTLE

A semi-aquatic freshwater turtle whose head and neck when extended are longer than its shell. The limbs are jointed and webbed, with 4 claws on the forefeet and distinct ankle joints. The nostrils are at the tip of the snout, enabling it to breathe while submerged, and it has a distinctive white ring around the iris. The shell is broad and wider towards the rear. It is fawn to black above with black-edged plates, and white to cream below, often stained brown by tannin in the water. Hatchlings have black shells with orange markings. **Behaviour** This turtle spends most of its life in the water creeping around the shallow edges and the floor of its waterhole searching for prey. It approaches victims holding its neck sideways close to the shell, and strikes with the mouth open, sucking in the prey which is ground between hard plates on the jaws. Large animals are shredded and pulled apart with the claws. In late summer, especially after rain, it may travel up to 2 km overland looking for a waterhole with food or a place to nest. In floods it is often found sleeping on high ground beneath fallen timber. In droughts it buries itself in mud, under litter or loose soil, and awaits the rain. They sometimes bask in the sun and southern populations hibernate in winter, digging into the mud at the bottom of a waterhole, extracting enough oxygen from the water through their skin. If alarmed they retract their head and neck under the shell, snap at the aggressor and squirt a foul-smelling, viscous fluid from glands just above the legs. **Development** Females lay up to 3 clutches of 8–24 brittle-shelled eggs from late spring to early winter. They are deposited in a hole dug about 120 mm deep into soil in an open site up to 500 m from a water body. The eggs hatch 3–6 months later and the hatchlings dig their way out, usually after or during rain, and rush to the water under cover of darkness. They may live to 36 years. **Diet** Any small animal including fish, molluscs, tadpoles, frogs, snails, crustaceans, carrion and aquatic vegetation. **Habitat** Freshwater ponds, lakes and streams.

SHELL LENGTH **To 300 mm** | STATUS **Low risk.**

FAMILY **CHELIDAE**   SPECIES *Chelodina oblonga*

OBLONG TURTLE

A freshwater turtle with a very long, thick neck and a long, flattened head. The head and neck are almost as long as the shell when fully extended. The limbs are jointed and webbed, not paddle-shaped, and the forefeet have 4 claws. The nostrils are at the tip of the snout, enabling it to breathe while submerged. The shell is oblong to narrow oval in shape, grey, olive-brown, dark brown to black above, sometimes with darker flecks, and white below. Juveniles are pear-shaped.

Behaviour This turtle is active year round and is often seen basking on logs or banks in the winter months. It catches prey by lunging forward with its mouth wide open. It grinds its catch between hard plates on its jaws and shreds larger animals with its front claws. Waterbirds are held under water until they drown and then shredded by a number of turtles. It regularly migrates overland to find new waterways or nesting sites, often travelling large distances and walking quite fast with its head and neck held straight out in front. It travels at night over open ground, and during the day if there is enough dense vegetation to hide in. In droughts it buries itself in the drying mud for 5–6 months, or migrates to another waterhole. If threatened it withdraws its head and neck under the front of its shell, folding them horizontally, and secretes a foul-smelling viscous fluid from glands in its legs. **Development** They mate in the water in winter and spring, and females lay 2–3 clutches of 3–16 brittle-shelled eggs from September to January. They are deposited in a hole about 150 mm deep, dug into soft sand or moist soil in relatively open sites up to 300 m from the water. The eggs hatch 7–10 months later. The hatchlings dig their way out and move to the water. **Diet** Small fish, crayfish, insects, molluscs, tadpoles, frogs, birds. Hatchlings eat mosquito larvae, water fleas, small fish and plants. **Habitat** Permanent freshwater rivers, lakes, ponds, dams and swamps.

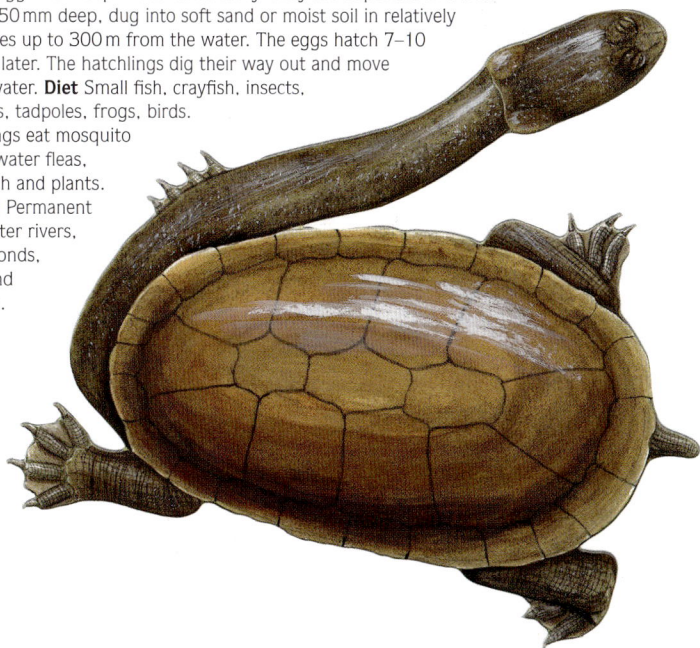

SHELL LENGTH **To 500 mm**   STATUS **Low risk.**

FAMILY **CHELIDAE**   SPECIES *Chelodina rugosa*

NORTHERN LONG-NECKED TURTLE

A tropical freshwater turtle with a broad, flattened head and a very long neck. When extended the head and neck are longer than the shell. The limbs are jointed with webbed feet instead of flippers, and the front feet have 4 claws. The nostrils are at the tip of the snout, allowing it to breathe while submerged. The shell is broadly oval and wider at the rear; it is dark brown to black above, often with darker flecks and blotches, and whitish below. **Behaviour** This turtle is a very efficient predator. It stalks fish among tree roots, floating vegetation and in shallow water near steep banks, and strikes with great speed, lunging with its long neck and sucking the victim into its mouth as it strikes. It also conceals itself in sediment or among plants with its head partly exposed, and ambushes passing prey. It can survive droughts by burying itself in the drying mud for 5–6 months, or migrating to another waterhole. It can tolerate saline water in coastal swamps, and large numbers gather in permanent waterholes at the end of the dry season. In the wet season it often travels to temporary waterholes on overcast days. If threatened it withdraws its head and neck under its shell. Unlike other members of this genus it does not produce a foul-smelling fluid when alarmed. **Development** They mate in the water, and females lay up to 4 clutches of 7–20 brittle-shelled eggs at the end of the wet season through the dry season, from March to September. The eggs are laid in moist soil in a hole dug in a bank by the water's edge where they are incubated by the heat of the sun, or in a nest dug in shallow water near the edge of the water body. They hatch at the beginning of the next wet season, from December to January. **Diet** Fish, crustaceans, tadpoles, frogs, aquatic insects, and rarely other reptiles. **Habitat** Tropical waterholes, swamps, billabongs, dams and slow-flowing rivers.

SHELL LENGTH **To 400 mm**   STATUS **Low risk.**

FAMILY **CHELIDAE**	SPECIES *Elseya dentata*

NORTHERN SNAPPING TURTLE

A tropical freshwater turtle with conspicuous, soft, rounded, wart-like projections on the back of its neck. It has a moderately long neck, although when extended the head and neck are much shorter than the shell. It has a horny casque over its head, and the upper jaw has a sharp, horned beak. Older adults often develop huge, muscular heads (known as 'boof-heads'). It has nostrils on the tip of its snout so that it can breathe while partly submerged. The limbs are jointed and webbed with 5 claws on the forefeet. The shell is broadly oval and wider to the rear, and the plate at the edge of the shell at the back of the neck is much narrower than the plates on either side of it. In young turtles the shell has a serrated rear edge. The shell is brown to dark brown above, often with scattered darker blotches, and whitish below. Females are larger than males. **Behaviour** This very slow-moving turtle has a varied diet and snaps at almost anything that comes within reach. It uses its powerful jaws and sharp, horned beak to slash through the flesh and bones of prey. Large food items are grasped in the jaws and shredded by the front claws. As the waterholes evaporate in the dry season it seeks a moist place by digging into the mud, hiding in a soil crack, hole, or among vegetation, and becomes inactive until the next rains. Large adults are often aggressive when molested, and may inflict a serious bite if handled. If the turtle is threatened it retracts its head and neck under its shell, folding them horizontally. **Development** They mate in the water and females lay a clutch of 5–16 elongated brittle-shelled eggs, about 55 mm long, in the dry season, from February to July. The eggs are deposited in a hole dug into moist soil in a sandy bank by the water's edge, covered with sand and incubated by the heat of the sun. The hatchlings emerge at the beginning of the wet season. **Diet** Mostly flowers, bark, leaves, figs, pandanus and other tropical fruits that fall into the water, with some fish, crustaceans, mussels and carrion. **Habitat** Freshwater lagoons, billabongs and permanent rivers.

SHELL LENGTH **To 450 mm**	STATUS **Low risk.**

FAMILY **CHELIDAE**   SPECIES *Wollumbinia latisternum*

SAW-SHELLED TURTLE

A widespread freshwater turtle with a horny shield covering
the top and the upper sides of its head, and prominent,
soft, spine-like projections on the back of its neck. It has
a moderately long neck, although when fully extended the
head and neck are much shorter than the shell. The limbs
are jointed and webbed, and the front feet have 5 claws. The
shell is broadly oval and wider at the rear with a serrated
edge. The plate at the edge of the shell behind the neck is as
wide as or wider than the plates on either side of it. The shell is brown
to dark brown above and whitish below. A sharply defined pale line runs
along the jaws and the side of the face. The nostrils are at the tip of
the snout, allowing it to breathe while partly submerged. Females are
much larger than males. **Behaviour** This slow-moving, river-dwelling
turtle feeds in the water and will snap at any passing object, grasping
and grinding prey in its jaws and shredding larger items with its claws.
The head and neck can be completely retracted under the front edge of
the shell by folding horizontally if the turtle is threatened. It also emits
a strong, foul-smelling odour from glands in its legs. It is generally
shy and placid, but may bite savagely in self defence if handled.
Development They mate in the water, and females lay several clutches
of 9–17 elongated, brittle-shelled eggs from September to January. They
are deposited in a hole dug into moist soil high up in the river bank, and
up to 53 eggs may be laid in a season. The eggs incubate in the heat
of the sun and hatch about 2 months later, before the onset of winter.
Diet Fish, molluscs, crustaceans, tadpoles, frogs, aquatic insects, water
weeds and fruits that fall into the water. **Habitat** Most watercourses,
especially medium, large and fast-flowing rivers.

SHELL LENGTH **To 280 mm**   STATUS **Low risk.**

FAMILY **CHELIDAE**   SPECIES *Emydura macquarii krefftii*

KREFFT'S RIVER TURTLE

A subspecies of *E. macquarii*, this widespread freshwater turtle
has a relatively short neck, jointed limbs with webbed feet
and 5 claws on the forefeet. Older individuals commonly
develop a large, muscular head. The shell is broadly oval and
usually is not flared at the rear. In many juveniles the rear
edge of the shell is serrated. The shell is light to dark brown
above, often with darker blotches, and white or cream tinged
with bluish-green below. A pale yellow stripe runs behind the eye
and along the side of the head. The nostrils are at the tip of the snout,
allowing it to breathe while partly submerged. Females are larger than
males. **Behaviour** This river and pond-dwelling turtle is often seen basking
in the sun, or resting on exposed logs and rocks in and around its
waterhole. Large numbers often congregate around a waterhole, lying
with their limbs and neck extended. They swim among logs and walk along
the bottom of the water body, and shelter beneath logs and other
vegetation. If threatened it withdraws its head and neck under the front of
its shell, folding it horizontally, and secretes a smelly but inoffensive
viscous fluid from glands in its legs. **Development** They mate in autumn
and winter, and females lay up to 5 clutches of 4–20 brittle-shelled eggs in
spring or early summer. The eggs are deposited in a nest dug among
vegetation or next to a rock or log on the bank of their waterway. They
hatch in summer after an incubation period of 2–3 months. **Diet** Adults
eat mostly freshwater sponges, insects and their larvae, crayfish, shrimps,
mussels, snails and the shoots of aquatic plants. Hatchlings and juveniles
feed mainly on aquatic insects. **Habitat** Lives in the larger rivers and their
associated permanent billabongs, swamps and larger waterholes.

SHELL LENGTH **To 340 mm**   STATUS **Low risk.**

| FAMILY **CHELIDAE** | SPECIES *Emydura macquarii macquarii* |

MACQUARIE RIVER TURTLE

A subspecies of *E. macquarii*, this widespread freshwater turtle has a relatively short neck that when combined with its small head is less than the length of its shell. It has jointed limbs with webbed feet and 5 claws on the forefeet. Like other freshwater turtles it has nostrils at the tip of its snout so that it can breathe while almost completely submerged. The shell is broadly oval, commonly flared at the rear with a serrated rear edge in juveniles. It varies from fawn to light brown above and is creamish-white to pale-yellow below. A pale yellowish stripe runs along the lower jaw to the side of the neck. **Behaviour** This turtle can live at the highest altitude of any Australian turtle (up to 420 m). It feeds mainly by day. In cool weather it basks in the sun on shallow banks and partly submerged logs, lying with its limbs and neck extended to absorb as much heat as possible. Adults also bask in the layers of warm water on the surface of still, deeper water. When threatened it withdraws its head and neck under the front of its shell. Due to the high rate of evaporative water loss, it does not migrate overland in dry times, and is confined to permanent waterways. **Development** They mate in spring and females lay up to 3 clutches of 6–30 elongated, brittle-shelled eggs, about 36 mm long, from mid September to early January. The eggs are laid after moderately heavy rain, on warm, overcast or rainy evenings. They are deposited in a hole 150–200 mm deep, dug high up into the bank in an open site 2–40 m from the water. They hatch in February after incubating for 2–3 months, and dig their way to the surface. **Diet** Fish, tadpoles, crustaceans, molluscs and aquatic vegetation. Young turtles feed mainly on aquatic insects and their larvae. **Habitat** Prefers deep, permanent still waterbodies in larger rivers and their associated lakes and billabongs, avoiding shallow water.

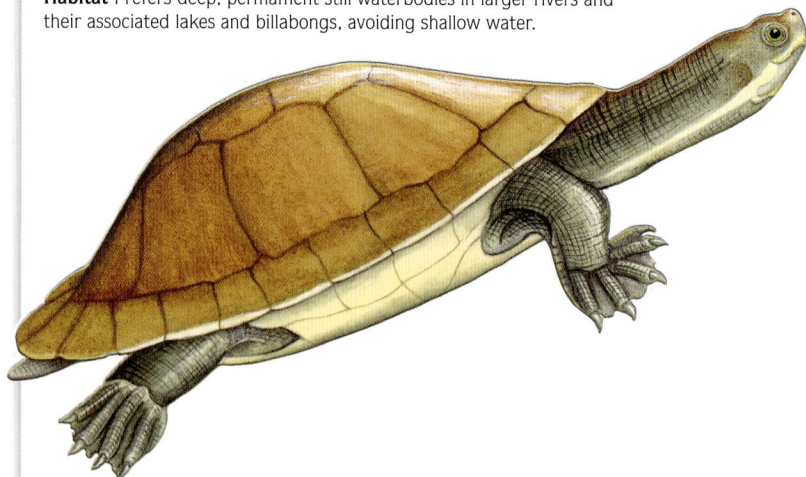

| SHELL LENGTH **To 300 mm** | STATUS **Low risk.** |

FAMILY **CHELIDAE**   SPECIES *Emydura victoriae*

NORTH-WEST RED-FACED TURTLE

A tropical freshwater turtle with a relatively short neck. The roof of the mouth is expanded to form a crushing plate, and many adults have grossly enlarged, muscular heads (known as 'boof-heads'). The limbs are jointed and have webbed feet and 5 claws on the forefeet. The shell is broadly oval, wider at the rear, brown to blackish-brown above with darker flecks and blotches, and whitish below. A bright red stripe, fading with age, runs from the eye and along the side of the head to the neck. The iris is an unbroken ring without leading or trailing dark spots. It has nostrils on the tip of its snout, enabling it to breathe while almost completely submerged. **Behaviour** This turtle is often seen basking in the sun on exposed logs and rocks in and around its waterholes, lying with its limbs and neck extended. It prefers smaller waterholes where it can hide from predators under overhanging banks. If threatened it withdraws its head and neck under the front of its shell, folding them horizontally, and secretes a smelly but inoffensive viscous fluid from musk glands in its legs. **Development** Females lay a clutch of around 16 brittle-shelled eggs between August and November. They are deposited in a hole dug close to the water's edge, or under leaf litter among the roots of a pandanus tree. **Diet** Pandanus fruits, crustaceans, mussels and aquatic insect larvae. Hatchlings and juveniles are carnivorous and feed mainly on aquatic insects. **Habitat** Small permanent waterholes, billabongs and the upper reaches of rivers.

SHELL LENGTH **To 300 mm**   STATUS **Low risk.**

Index